Animal Locomotion

Oxford Animal Biology Series

Editors

Professor **Pat Willmer** is in the School of Biology at the University of St Andrews. Dr **David Norman** is Director of the Sedgwick Museum at the University of Cambridge.

Advisers

The role of the advisers is to provide an international panel to help suggest titles and authors, to ensure individual countries' teaching needs are met, and to act as referees.

The Oxford Animal Biology Series publishes attractive supplementary textbooks in comparative animal biology for students and professional researchers in the biological sciences, adopting a lively, integrated approach. The Series has two distinguishing features: first, book topics address common themes that transcend taxonomy, and are illustrated with examples from throughout the animal kingdom; secondly, chapter contents are chosen to match existing and proposed courses and syllabuses, carefully taking into account the depth of coverage required. Further reading sections, consisting mainly of review articles and books, guide the reader into the more detailed research literature. The Series is international in scope, both in terms of the species used as examples and in the references to scientific work.

Animal
Locomotion

Andrew A. Biewener

Charles P. Lyman Professor of Biology
Director, Concord Field Station, Harvard University

UNIVERSITY PRESS

This book has been printed digitally and produced in a standard specification
in order to ensure its continuing availability

OXFORD
UNIVERSITY PRESS

Great Clarendon Street, Oxford OX2 6DP

Oxford University Press is a department of the University of Oxford.
It furthers the University's objective of excellence in research, scholarship,
and education by publishing worldwide in

Oxford New York

Auckland Cape Town Dar es Salaam Hong Kong Karachi
Kuala Lumpur Madrid Melbourne Mexico City Nairobi
New Delhi Shanghai Taipei Toronto
With offices in
Argentina Austria Brazil Chile Czech Republic France Greece
Guatemala Hungary Italy Japan South Korea Poland Portugal
Singapore Switzerland Thailand Turkey Ukraine Vietnam

Oxford is a registered trade mark of Oxford University Press
in the UK and in certain other countries

Published in the United States
by Oxford University Press Inc., New York

ISBN 978-0-19-850022-3

Printed and bound by CPI Antony Rowe, Eastbourne

Preface

In writing this book my immediate aim was to provide a synthesis of general physical, physiological, and biomechanical principles that underlie the many ways in which animals move. An understanding and full appreciation of animal locomotion requires the integration of these principles. At the same time, I sought to examine a reasonably broad scope of the diversity of the many fascinating mechanisms that animals have evolved to move. Necessarily, many readers will find that certain interesting topics and examples have been left out. Nevertheless, my hope is that I have managed to cover a range of topics and examples that represent the most salient features of animal movement that have been evolved across the animal kingdom. Over the years, other fine books have been written covering various aspects of animal locomotion. These texts, together with my own research and interactions with many valued colleagues, have shaped my thinking and strongly influenced what is contained here. Often such texts are focused on a narrower range of animal locomotion, or have emphasized a particular aspect of animal movement. Animal locomotion is indeed a broad topic. My goal therefore has been to present, as broadly as possible and within a reasonable amount of space, a discussion of animal locomotion that is accessible to undergraduates, yet also of value to more advanced graduate students and professionals. Toward this end, I provide the necessary introductory foundation that will allow a more in-depth understanding of the physical biology and physiology of animal movement. In so doing, my hope is that this book may also serve as a useful springboard for further reading of more advanced texts that cover certain aspects of animal movement introduced here in more detail.

This book attempts to provide a current synthesis of much of animal locomotion as we know it today. Because science and our knowledge of the biological world continually evolve, it is my hope that the ongoing discovery of new ideas and mechanisms that underlie any field also emerges in this book. In keeping with this, I have attempted to identify those directions in which future research may lead us to a greater appreciation and understanding of animal locomotion and its broader importance to the biology of the organisms that we study and love.

Several themes run through this book. The first is that by comparing the modes and mechanisms by which animals have evolved the capacity for movement, we can come to understand better the common principles that underlie each mode of locomotion. A second is that size matters. One of the most amazing aspects of biology to me is the enormous spatial and temporal scale over which organisms and biological processes operate. Within each mode of locomotion, animals have evolved successful designs and mechanisms for contending with the physical properties and forces imposed on them by their environment. Understanding constraints of scale that underlie locomotor mechanisms is essential to appreciating how these mechanisms have evolved and how they operate. A third theme is the underlying importance of taking an integrative and comparative evolutionary approach in the study of biology. Organisms share much in common. Much of their molecular and cellular machinery is the same. They also must contend with similar physical properties of their environment. Consequently, an integrative approach to organismal function that spans multiple levels of biological organization provides a far better understanding of animal locomotion. By comparing across species, common principles of design emerge. Such comparisons also highlight how certain organisms may differ, pointing to new adaptive strategies that have evolved for movement in diverse environments. Finally, because convergence upon common designs or the generation of new designs results from historical processes governed by natural selection, it is also important that we ask how and why these designs have evolved.

I have many people to thank in writing this book. First, I thank my many students and colleagues with whom I have shared the fascination and love of animal movement, physiology and biomechanics. The interactions that have come from our work and discussions together are the best part of science. Individually, I thank George Lauder, Bob Shadwick, and Gary Gillis for their helpful comments and feedback on Chapter 4; Bret Tobalske, Jim Usherwood, and Ty Hedrick for our many discussions and comments on Chapter 5; Bob Full, Tom Roberts, and Peter Weyand for discussions (and debates) of topics covered in Chapters 3, 8 and 9; and finally, Michael Dickinson for feedback and discussions of topics covered in Chapters 5 and 10. I also have many people to thank for help with figures that are used in the book, many which are reproduced from copyrighted material published in their journal papers and books. I am particularly grateful for these, as completing the figures for the book proved to be a much more difficult and time-consuming task than I had anticipated.

I am also grateful for the patience of my editors, who put up with the overly long time that it took for me to compete this book. Finally, I express my lasting appreciation to my wife, Gayle, whose support and patience over the years has been marvelous, particularly owing to the many distractions she faced when I was writing this book and the time that it added to an already overcommitted schedule.

This book is dedicated to Dick Taylor, whose unbounded enthusiasm for comparative physiology and love of animal locomotion were an inspiration to me, as well as to so many others in our field.

Contents

1 | Physical and biological properties and principles related to animal movement

1.1 Why move?

One of the defining characteristics of animals is their movement. Whether the animal is single celled or is much larger and has appendages, active foraging for food sources, movement to avoid a stressful environment, active pursuit of prey and finding a mate are all behaviors that animals engage in by means of locomotor movement. Beyond its fundamental biological importance, our fascination with animal movement often derives from observing the beauty, grace and sheer athleticism of animals as they move. What aspects of flight do darting hummingbirds and bumblebees share in common? How do they differ from a soaring petrel? What are common patterns of design shared by a racing antelope, a scurrying lizard or running cockroach, and in what ways do they differ? The grand scale of size and diversity within biology adds to the impressive range of locomotor mechanisms that animals have evolved. Yet, despite this amazing diversity, surprisingly common principles and biological components underlie most of these mechanisms. Therefore the study of animal locomotion depends on an understanding of these physical principles and the properties of the media which influence how animals move. This, in turn, helps us to understand why certain biological structures have evolved for movement within particular physical environments.

This book is about how animals move. It addresses basic physical principles and properties of the media in which animals move, seeking to explain the design and locomotor function of animals within these media. It also attempts to capture much of the amazing diversity of animal design and movement. Much of this diversity arises from the enormous scale of size within biology, ranging from single-celled organisms to the largest whales (10^{15} orders of magnitude in mass), and the breadth of environments that animals inhabit. Therefore size and environment are important underlying themes of the book.

1

At the same time, in the attempt to discuss the basic principles of design and function which underlie locomotor diversity, certain animals and loco-motor devices inescapably must be left out. It is my hope that the richness, challenge and excitement of studying and understanding animal locomotion springs forth from the topics that are covered in this book.

1.2   Environmental media

Land, air and water constitute the physical world of organisms. To a large extent, the properties of these media dictate the locomotor mechanisms evolved by the animals which live within them. For animals that move on land and fly, the properties of the air and gravity dominate their physical world. However, gravity is of little concern for aquatic animals. In addition, air and water play an important role as respiratory media and hence affect locomotor design in terms of how energy is supplied for powering and sustaining movement. The capacity to move between physical environments is also important to many animals. This is the case for flying animals which must also be capable of support and movement on land, as well as for animals living near the shore, in which air and water movement is also often linked.

1.2.1   Physical properties of media

Table 1.1 lists several important physical properties of air and water that influ-ence the way in which animals are designed and the way they move. Air and water can both be thought of as fluids. Fluid movement past the body of organisms is fundamental to nearly all forms of animal locomotion (except burrowing). Most important is the difference between the densities of water and air—the density of water exceeds that of air by more than a factor of 800. Even though aerial flight and aquatic locomotion depend on the same fluid mechanical principles, the difference between the densities of these two media has significant implications for the requirements of these two modes

Table 1.1 Physical properties of media

Physical property	Air	Water	Ratio
Density (g/cm^3)	0.0012	1.000	830
at 25 °C		1.02 (seawater)	
Dynamic viscosity	18×10^{-6}	1×10^{-3}	55
(Pa s = Ns/m) at 20 °C			
Oxygen content (ml O_2/l)	209	7	30
Heat capacity			
(cal/l °C)	0.31	1000	3200
(J/l °C)	1.30	4184	3200

of locomotion, as well as for the terrestrial locomotion of land animals. The difference in viscosity, although smaller in magnitude, also has an important influence on how fluid flows past an organism as it moves. The differences in oxygen content and heat capacity of air compared with water influence the locomotor design of animals in a more indirect way, by affecting their thermal and respiratory functions. As will be seen, the locomotor capacity and strategy of animals depends on their physiological capacity to deliver oxygen to their tissues, specifically their muscles, and to generate metabolic energy in the form of adenosine triphosphate (ATP). These two functions are most influenced by temperature and the availability of oxygen supply from the environment, although differences in density also greatly affect the respiratory design and function of animals.

1.2.2 Impact of physical media on locomotor function

Because of its much lower density and viscosity, air imposes proportionately smaller resistive (drag) forces for flying and terrestrial animals than water does for aquatic animals. Therefore the main problem for terrestrial animals lies in overcoming mass-related gravitational forces as they move. The low density of air also means that flying animals must generate sufficient aerodynamic force (lift) to support their weight, in addition to aerodynamic thrust to overcome drag associated with moving in a forward direction. Aquatic animals, on the other hand, need not worry much about supporting their weight because the density of their bodies nearly matches that of water. Therefore most aquatic animals are neutrally, or slightly negatively, buoyant in water. However, the higher density of water means that drag poses a formidable obstacle to their movement. Consequently, drag reduction is critical, particularly for animals of moderate to large size.

Differences in the oxygen content and heat capacity of the two media also affect the activity levels and locomotor strategies of animals. The greater oxygen content of air generally affords higher levels and broader strategies of activity for flying and running animals than for swimming animals. The higher heat capacity of water further constrains the locomotor capacities of swimming animals by making it more difficult for them to maintain a warmer body temperature than their surrounding environment. Having said this, however, there are many exceptions to these general rules. Aquatic and cold-acclimatized animals have evolved, and can adaptively express, metabolic enzymes which work well at low temperatures, enabling them to compensate for a colder environment. In addition, differing metabolic pathways for energy production afford animals varied locomotor strategies for daily activity which enable equally successful performance compared with that achieved by warmer animals.

1.3 Physics and energetics of movement

Animals move by exerting forces (measured in SI units of Newtons (N)) on their external environment, whether it be a solid substrate or a fluid (air or water). By Newton's First Law

$$F = ma \qquad (1.1)$$

where m is the mass (in kilograms (kg)) of the body moved and a is its acceleration (in m/s^2). Therefore an animal's weight is the force produced by the acceleration g due to the Earth's gravity acting on its mass m. To move its body, an animal must do work

$$W = Fd \qquad (1.2)$$

where d is the distance (in metres (m)) that the animal's body moves as a result of the net force acting on it, in reaction to the forces that the animal transmits to the environment. Work (in joules (J)) represents the mechanical *energy* required to move the animal's body. The amount of mechanical energy required to move per unit time

$$P = W/t = Fd/t = Fv \qquad (1.3)$$

represents the mechanical power (in watts (W)) of locomotion, and thus can be related to the forces that an animal exerts as it moves a given distance per unit time, or the velocity v of its movement.

The energetic efficiency of an animal's movement can be calculated by comparing the metabolic energy consumed (energy input) with the mechanical work (energy output) done over a given period of time:

$$\text{efficiency} = \text{energy out/energy in} \qquad (1.4)$$

$$= \text{work/metabolic energy} \qquad (1.5)$$

or, equivalently, the mechanical power output versus the metabolic power input (P_{out}/P_{in}). Typically, locomotor efficiencies are determined by comparing the oxygen consumption of an animal with the mechanical work performed over an integral number of strides. Because all animals must ultimately balance their energy needs by means of aerobic (oxygen-dependent) metabolism, measurements of oxygen consumption are commonly used to assess the energy supply of ATP needed for sustainable locomotor activity. Typically, a value of 20.1 kJ/l O$_2$ is used. This value assumes that ATP is produced by means of aerobic glycolysis (the breakdown of glycogen into glucose and its transformation via glycolysis and the Krebs cycle into ATP production within the mitochondria by electron transport and oxidative phosphorylation).

1.4 Biomechanics of locomotor support

The forces required for locomotion are typically generated by muscles, or by motor proteins within cellular animals, and are transmitted to the external environment by means of a skeleton. These forces cause deformations in the structures that transmit them. The ability of a structure to resist deformation when subjected to a given force is a measure of its stiffness and is the slope of a structure's force–length relationship (Fig. 1.1(a)). Linearly elastic structures are defined as having a linear force–length relationship, typical of a simple spring that is stretched. Although linear elasticity is easier to analyze, many biological structures exhibit non-linear elasticity. Because larger structures can support larger forces, engineers commonly normalize for differences in the size of structures by dividing the force acting on a structure by the its cross-sectional area A (Fig. 1.1(b) and 1.1(c)). When normalized in this way, a force is defined as a stress (denoted by Greek sigma)

$$\sigma = F/A. \tag{1.6}$$

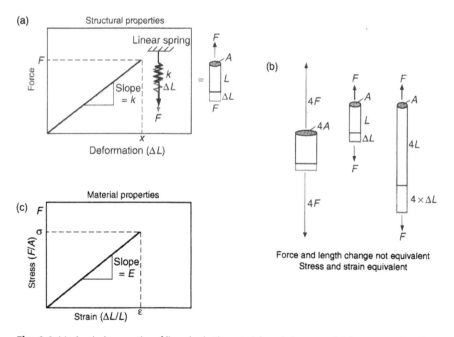

Fig. 1.1 Mechanical properties of linearly elastic materials and structures: (a) force versus length change; (b) force and length change vary in relation to a structure's size; (c) stress versus strain normalize for differences in size and thus reflect the material properties of a structure. Similar to the slope of force versus deformation (a), which represents the spring stiffness k of a structure, the slope of stress versus strain (c) represents the elastic modulus E of the material.

The common units of stress relevant to the musculoskeletal systems of animals during locomotion are N/mm^2 ($= 10^6 \, N/m^2 = 1 \, MN/m^2$, or 1 MPa) or N/cm^2 ($= 10 \, kN/m^2$, or 10 kPa). Because these units of stress may be new to most readers, and also counterintuitive, a useful example is that the weight of an apple (about 1 N, certainly an apposite definition of a Newton!) balanced on the end of a toothpick (of cross-sectional area 1 mm^2) exerts a stress of 1 MPa. Whereas forces act *on* structures, stresses can be thought of as being transmitted *through* the structure. Large structures also undergo larger deformations than smaller structures. Once again, in order to account for differences in size, deformations or changes in length are normalized by dividing by the structure's resting (unloaded) length (Figs 1.1(b) and 1.1(c)), and are defined as a strain (denoted by Greek epsilon):

$$\varepsilon = \Delta L / L. \tag{1.7}$$

As engineering terms, therefore, stress and strain have quite distinct meanings. However, whereas strain represents the real physical deformation of a structure in response to being loaded, the stress acting within the material represents a conceptualization of the intensity of force transmission. Finally, by considering the size-independent properties of a material, stress and strain have an equivalent relationship to that of force and length (Fig. 1.1(c)), in which the stiffness of the material is the slope of the stress–strain relation, and is defined as the elastic modulus (also known as Young's modulus)

$$E = \sigma / \varepsilon. \tag{1.8}$$

The force causing a structure to break (Fig. 1.1(a)) can be related to the strength, or maximum stress (Fig. 1.1(c)), that the structure can bear before failing. This also defines the strain at failure. The area under the force–length curve represents the amount of energy ($(F \times d)/2$ for linearly elastic elements) that is absorbed by a structure when it is loaded (Fig. 1.2(a)). Similarly, the area under the stress–strain curve represents the amount of strain energy absorbed *per unit volume* of material ($U^* = \sigma\varepsilon/2$) (Fig. 1.2(b)). If a structure is unloaded before breaking, this energy can be recovered elastically (much like a rubber ball or an elastic band) and may be used to do work. Elastic structures which can be used to store and recover elastic strain energy, and thus reduce the metabolic cost of locomotion, exist within animals.

The elastic modulus and the energy absorbed before failure define whether a material is 'rigid' or 'compliant' and 'brittle' rather than 'tough'. Rigid materials deform little when loaded and have a high elastic modulus, whereas compliant materials undergo considerable strain for a given load and have a low modulus. Tough materials absorb considerable elastic strain energy before failing, whereas brittle materials, such as glass, absorb very little (Fig. 1.2(c)). Generally, tough materials are not as rigid as brittle materials,

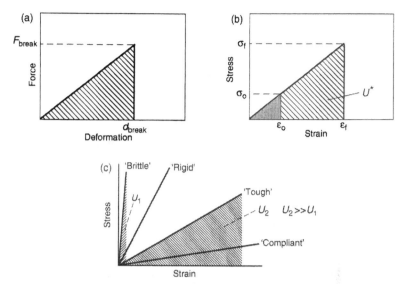

Fig. 1.2 (a) The energy absorbed by a structure when loaded represents the area under its force–deformation curve (for linearly elastic structures this is $(F \times \Delta L)/2$). (b) Similarly, the area under a material's stress–strain curve represents the energy absorbed per unit volume. Typically, the maximum operating stress σ_0 or strain ε_0 of a material is much less than its failure stress σ_f or failure strain ε_f. The ratio of a material's failure stress to its operating stress (σ_f/σ_0) is often used to define the safety factor of a material or a structure. (c) A comparison of the stress–strain curves for various types of materials.

i.e. they have a lower elastic modulus. On the other hand, although brittle materials may have a high failure strength and elastic modulus, they often fail relatively easily, especially when subjected to impact loads. Consequently, the amount of energy absorbed to failure is a measure of the material's 'toughness'. Because biological structures are often subjected to dynamic loads, their ability to absorb strain energy is often the critical factor determining whether they break. In general, most biomaterials have evolved designs which enable them to be tough, so that they can absorb a considerable amount of energy before breaking. As a result, rigid biomaterials exhibit a stress–strain relationship intermediate between brittle and compliant materials (Fig. 1.2(c)).

1.4.1 Modes of loading

The mechanical loading of support structures typically consists of four types of loads: axial tension, axial compression, bending, and torsion (Fig. 1.3). When subjected to axial loads, the stress developed depends only on the structure's cross-sectional area relative to the magnitude of the applied load. Tension is

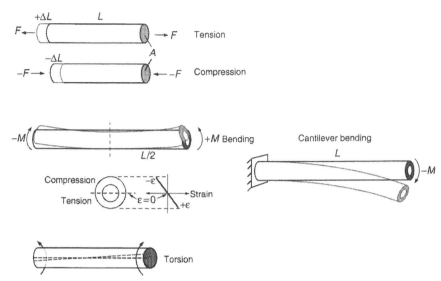

Fig. 1.3 Comparison of the main modes of loading to which structures are often subjected. Whereas long-axis tension and compression result in uniform stress or strain distributions across the cross-section of a beam, bending and torsion produce non-uniform gradients of stress and strain. In the case of bending, maximum tension and compression occur on opposing surfaces with a plane of zero strain across the beam's midsection (neutral plane of bending). In torsion, the location of zero strain is an axis running through the center of the beam. Typically, bending and torsion produce much greater strains, with the possibility of failure, than when a structure is loaded in tension or compression.

defined as an axial load which tends to elongate a structure, whereas compression is defined as a load which shortens the structure along a given axis. When subjected to bending, both tensile and compressive stresses act within a structure (Fig. 1.3(b)). Compression occurs on the concave surface and tension on the convex surface, with the greatest stress acting at the surfaces in the plane of bending. Consequently, there is a gradient of stress (and strain) from maximum compression on one surface to maximum tension on the opposite surface (Fig. 1.3(c)). This means that there is a neutral plane midway through the structure's cross-section where stress and strain are zero. If a structure is subjected to bending and axial compression or tension, this will cause a shift in the neutral plane, displacing it from the midpoint of the section. Unlike axial compression or tension, stresses due to bending depend on the shape of the cross-section as well as its size. This is because material located near the neutral plane of bending experiences lower stresses.

Consequently, beam-like elements with hollow, rather than solid, cross-sections provide much better resistance to bending for a given weight (Fig. 1.3(c)). This is why bicycle frames and many other structures (tent poles) are constructed with a tubular design. For the ambitious reader, the mechanical basis of this shape factor, termed the second moment of area, is provided in engineering texts (Beer and Johnston 1981; Wainwright *et al.* 1976). A symmetric tubular shape is favored when a range of bending loads in various directions are likely to act on the structure. Finally, the stresses developed in bending depend not only on the magnitude of the bending force (F_b), but instead on the bending moment ($F_b \times L/2$), in the case of bending stresses developed at a midpoint of a beam shown to the left and $F_b \times L$, in the case of stresses at the base of a beam subjected to cantilever bending shown to the right (Fig. 1.3(b)). This means that longer beams, such as the long bones of a vertebrate limb, are more likely to experience large bending-induced stresses than short elements, such as the vertebrae.

In addition to axial loading and bending, structures may also be loaded in torsion, which involves a rotational moment applied about the long-axis of a beam-like structure (Fig. 1.3(d)). As with bending, cross-sectional shape is again important to how well the torsion is resisted. In this case, shape depends on the distribution of material away from the neutral axis of torsion (for a circular beam this is the midpoint of the cross-section). In general, biological structures that are designed well to resist bending also provide effective resistance to torsional loads. This probably reflects the fact that strong twisting of the body about a structural member is not common in the natural movements of most animals. When it does happen, such as when one's ski binding fails to release in response to a twisting motion of the body when losing one's balance while skiing, the results are often unfortunate!

1.4.2   Safety factors

Like human engineered devices, biological structures achieve designs (by means of natural selection) that provide a 'factor of safety' in order to reduce their risk of failure. Safety factors are often defined in terms of the ratio of a structure's strength to the maximum stress that it is likely to experience over its lifetime of use (σ_f/σ_o) (Fig. 1.2(b)). Engineered structures typically are built to have safety factors as high as 10. This means that the maximum anticipated load would not exceed one-tenth of the structure's maximum load capacity. With a safety factor of 10, the chance of failure is quite low—a comforting fact when using an elevator to move up several floors of a tall building! With more stringent performance requirements (such as aircraft, which must minimize weight) or a lower cost for failure, the safety of a structure may be less. Safety factors can also be defined in terms of toughness and strain energy. However, because these are often much more difficult to measure, stress is used more often to define safety factors.

Biological safety factors favored by natural selection are generally less than the safety factors of engineered buildings and mechanical devices, more often ranging between 2 and 8. For example, in order for the tibia of a gazelle, which has a failure strength of 200 MPa, to maintain a safety factor of 4 during fast galloping or jumping, the size of its tibia and its manner of loading must ensure that the maximal stresses developed within the bone do not exceed 50 MPa during these locomotor activities. The lower safety factors of biological structures are probably due, in part, to the fact that animals must also pay a price for maintaining and transporting the mass of their tissues when they move. This cost is probably balanced against the benefit of a reduced risk of failure (Alexander 1981). Finally, it is likely that the failure of structures which would most reduce an animal's fitness, such as a primary limb bone rather than a distal phalanx or a feather shaft, would favor a higher safety factor. The relative incidence of bone fracture within thoroughbred race horses appears to provide evidence of this: fracture is highest in the distal hoof elements, lower in the proximal femur and humerus, and lowest in the vertebral column and skull (Currey 1981).

1.5   Scaling: the importance of size

Size is arguably one of the most important variables affecting the function and form of organisms. This is because changes in size that occur during growth and over evolutionary time impose changes in the relative dimensions of organisms which have important functional consequences. This is because many physiological processes and mechanical properties depend on key structural dimensions, such as surface area or thickness. When different-sized structures retain the same shape they are considered to scale isometrically, or to be 'geometrically similar'. For geometrically similar structures (Fig. 1.4), all linear dimensions scale in proportion to one another (i.e. lengths L ($L \propto L$ and $L \propto D$)) and areas A are proportional to L^2 or D^2, and to volume $V^{2/3}$.

Because of this, area-dependent processes change at a different rate than linear- or volume-dependent processes. This is important for both the physiological and mechanical functions of organisms. For example, the capacity of animals to sustain activity depends on their ability to transport oxygen and fuel substrates to the mitochondria inside their muscle fibers. Ultimately, this depends on the rate of diffusion across cellular and mitochondrial membranes, which in turn depends on the area of exchange surfaces. However, if the energy demand or work required to move the animal depends on its mass (or volume), this poses a scale-dependent constraint of energy supply relative to energy demand that is proportional to A/V or $V^{-1/3}$ ($M^{-1/3}$) for geometrically similar animals. In other words, a 100-fold increase in size can be expected to impose an almost fivefold reduction in an animal's capacity to

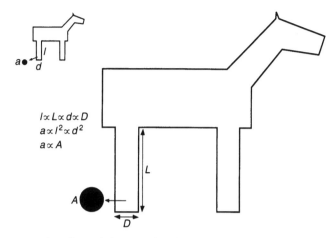

Fig. 1.4 Geometric scaling and change in the relative dimensions of different-sized organisms.

fuel its activity. This would mean that an animal's **mass-specific metabolism**, defined as the amount of energy that each gram of its tissue requires to meet its metabolic needs, would decrease fivefold due to the decrease in surface area relative to volume at larger size. The effect of size on energy metabolism associated with fueling locomotor activity is discussed at length in Chapter 9.

A similar area versus volume constraint operates with respect to mechanical support. This is because stresses depend on force per unit area , which means that stresses are likely to increase with size (again, for a 100-fold increase in mass, weight-related stresses can be expected to increase nearly fivefold). Unless the tissue strength of the skeleton increases in a similar fashion, the risk of failure may become exceedingly high. This means that animals built of similar materials must either evolve mechanisms for reducing the weight-related forces generated within their musculoskeletal systems or drastically restrict their performance.

To a certain extent animals may deviate from geometric similarity in order to compensate for the scale effects of size. When this happens distortions of shape, or **allometric** changes in structural and functional properties, occur with size. Allometric scaling might reflect, for example, either a relative shortening or lengthening of an element or its relative thickening or thinning for a given mass or area. When the scaling change is greater than that expected for isometry it is defined as positive allometry; when it is less than the isometric expectation, it is defined as negative allometry. Even moderate allometric scaling requires substantial distortions in shape when size changes over several orders of magnitude (Fig. 1.5). A good example of positive allometry is the scaling of mammalian lung surface area (Fig. 1.6), which was found to scale with a

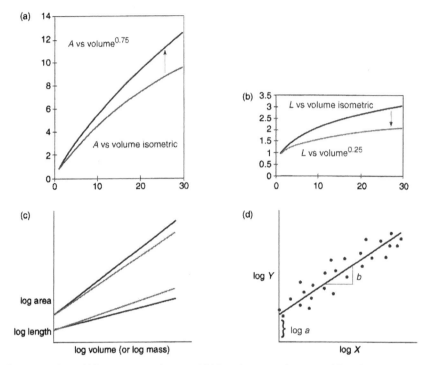

Fig. 1.5 Scaling of (a) area versus volume and (b) length versus volume on arithmetic coordinates and (c) on logarithmic coordinates. In each case, the isometric scaling pattern is the grey line and the allometric scaling pattern is the dark line (scaling of A versus D is positively allometric and scaling of L versus D is negatively allometric). (d) Sample scatter in biological data, with linear regression fit of the logarithmically transformed data to determine the scaling exponent (slope b) and coefficient (Y-intercept a) of the exponential relationship $Y = aX^b$.

slope of 0.92 when plotted on logarithmic axes. This indicates that the lungs of larger mammals are much more finely partitioned than would be expected if they were geometrically similar to the lungs of small mammals. The observed scaling of lung surface area also suggests a greater aerobic locomotor capacity than if the lungs of larger animals remained isometric in design (see Chapter 8). This provides an example in which the scaling of a key structural feature of the lungs which is important to diffusive gas exchange can be related to the metabolic demand for gas exchange. If, on the other hand, different-sized animals retain similar shape (i.e. scale close to geometric similarity), alternative mechanisms must often be evolved to compensate for functional constraints of size. We shall see how size affects locomotor mechanisms. Indeed, much of the locomotor diversity of animals reflects this fundamental aspect of their biology.

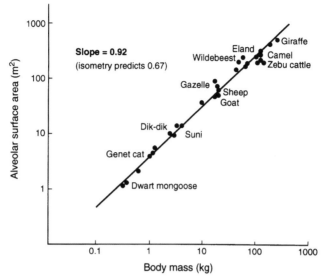

Fig. 1.6 Example of positive allometric scaling of lung alveolar surface area versus body mass in terrestrial mammals. Isometric or geometrically similar scaling would predict a slope of 2/3 (or 0.67); however, lung surface area is observed to scale with a slope of 0.92, indicating strong selection for pulmonary diffusion capacity so that oxygen uptake meets the increased metabolic demand for oxygen delivery at larger size. (Adapted from Gehr *et al.* 1981.)

1.5.1 Allometric equation

Geometric or allometric scaling of physiological functions and structural dimensions can be related to changes in size by the exponential equation

$$Y = aX^b \tag{1.9}$$

where b is termed the scaling exponent and a is the scaling coefficient relating changes in variable Y to changes in variable X. This equation can be linearized by means of logarithmic transformation

$$\log Y = \log a + b \log X \tag{1.10}$$

in which case, the scaling exponent becomes the slope b and $\log a$ is the Y-intercept of the line relating $\log Y$ to $\log X$ (Fig. 1.5(d)). Base-10 logarithms are generally used to linearize the exponential relationships describing the structural and physiological scaling of organisms. However, natural logarithms (ln, base e) are sometimes also used. The linear relationship described by eqn (1.10) has the benefit of allowing data to be plotted over several orders of magnitude and the use of regression methods for statistical evaluation of empirically determined relationships between two variables. Such **bivariate plots** commonly have scatter around the predicted scaling line, which provides a measure of the strength of the correlation between the two variables. Deviations from the observed scaling pattern may also provide important insight into how a particular species has modified its functional design.

1.6 Dimensions and units

It is important (and of practical use) to distinguish between dimensions and units in describing and analyzing the design of organisms. Dimensions represent the fundamental physical features of a variable. Variables such as force F are defined in terms of mass M and the mass acceleration a. Similarly, velocity is defined in terms of the dimensions length L per unit time T. The quantitative measure of dimensions is expressed in terms of units. Consequently, depending on the set of units used to measure them, variables will have different values. Units for force may be a dyne, a newton or a pound. Units of length may be inches, centimeters or meters. The Standard International (SI) system of units is now commonly adopted throughout the scientific community and is the system that will be used in this book. Because it is a metric system, forces are measured in newtons, lengths in meters, and velocities in meters per second (m/s).

All biological and physical variables can be defined in terms of three fundamental dimensions: length, mass and time. Several variables with their commonly used dimensions and fundamental dimensions are shown in Table 1.2. These dimensions provide a means for ensuring that equations are dimensionally correct (which is of equal, if not greater, importance than being quantitatively correct, as quantitative accuracy depends on dimensional accuracy). This, in turn, can often help one identify a key variable which may be missing from an equation which is found to be dimensionally incorrect.

Table 1.2

Variable	Common dimensions	Fundamental dimensions
Force	Mass and acceleration	MLT^{-2}
Velocity	Distance and time	LT^{-1} (equivalent in this case)
Work	Force and distance	ML^2T^{-2}
Stress	Force and area	$ML^{-1}T^{-2}$

2 | Muscles and skeletons: the building blocks of animal movement

Before considering various modes of animal locomotion, we shall first review the organization, physiology and biomechanical properties of muscles and skeletons. These are the elements upon which animal movement depends. An animal's musculoskeletal system encompasses the mechanical interaction of its muscles with various skeletal elements to transmit force for movement and support. As we shall see, muscles not only function as machines to do work by contracting to shorten while generating force, they can also operate as brakes to slow an animal or a movement of an appendage, or they may function as a strut to maintain the position of a joint and potentially facilitate elastic energy storage and recovery. Muscles in all animals share a similar basic organization, relying on the same protein machinery for generating force and movement. Therefore differences in muscle function depend more on the relative amounts of their underlying mechanical and energetic components, differences in their enzymatic properties, and differences in how they are activated by the nervous system. Muscles (and the motor proteins of cells) require a rigid skeletal system to transmit force for movement and support. In most animals, muscle force transmission is linked via series of jointed skeletal segments and levers. Thus the properties of the skeleton and the overall design of musculoskeletal organization are important aspects of how muscles transmit the forces that are necessary for animals to support themselves and move through their environment.

2.1 Muscles

All multicellular animals rely on striated skeletal muscles to power their locomotion. Certain unicellular organisms and individual cells also use proteins similar to those found in the skeletal muscles of larger animals as motors to alter their shape and move. These will be discussed in Chapter 6. In general,

15

muscles generate energy for movement by doing work in order to function as a biological motors. Muscles do this by exerting force F while shortening (change in length ΔL)—hence the term 'muscle contraction'. The product of force and length change equals work ($W = F \times \Delta L$), which is measured in joules (1 J = 1 N × 1 m). Muscles most commonly change length over distances of millimeters, so that the work they perform is given in millijoules (mJ). Work per unit time, in turn, equals the power ($P = F \times \Delta L/\Delta t$) produced by a muscle, which is measured in watts (1 W = 1 J/s). By definition, muscles produce positive power when they shorten (i.e. *decreases* in length are defined as being *positive*). However, as we shall see, muscles may also function to generate force with little or no change in length, in which case the contraction is referred to as being 'isometric'. Ideal isometric contractions result in zero work and power. Other muscles may lengthen as they generate force, thereby absorbing energy by doing 'negative work' (ΔL is defined as being negative in this case).

2.1.1 Molecular organization of muscle: mechanism of force generation and shortening

Why do muscles generate force by shortening? The muscles of all multicellular animals are comprised of two basic motor proteins: **actin** and **myosin**. These two proteins exist in a regular (and highly conserved) hexagonal lattice which constitutes the myofilaments of muscle cells (Fig. 2.1(a)). The elongate muscle cells form the muscle fibers. In nearly all muscles, the muscle cells are extremely long relative to their diameter (e.g. 20–40 mm long × 0.03 mm diameter in a dog). Even in much smaller molluscs and insects, muscle cells exhibit a similarly high aspect ratio (1–5 mm × 0.002 mm). Overlapping sets of actin (thin) and myosin (thick) filaments are arranged in series along the length of the muscle fiber in functional repeating units called **sarcomeres**. Each sarcomere is composed of two sets of actin filaments extending from either end (the Z-disk) of the sarcomere, which overlap by interdigitating with the myosin filaments extending from the sarcomere midline. The sarcomeres themselves are organized in series (joined together at neighboring Z-disks) as an intermediate myofibril which runs end to end within the muscle fiber. The regular spacing of thick and thin filaments as repeating sarcomeres along the length of the myofibrils gives skeletal and cardiac muscles a 'cross-striated' appearance under both light and scanning electron microscopes (Figs 2.1(a) and 2.1(b and c)). Consequently, these muscles are often referred to as striated muscle (in contrast with smooth muscle found in arteries, the gut, and elsewhere, which lacks this regular sarcomeric organization of myofibrils). Cross-striated vertebrate and invertebrate muscles should also be distinguished from the obliquely striated muscles of annelids and cephalopods (Hoyle 1983; Kier 1996), which are not discussed here.

The myosin filaments are comprised of a polymeric chain of myosin protein elements, each consisting of a heavy chain and two light chains which form a pair of globular domains at the 'head' end of the myosin. Each myosin head is flexible and capable of undergoing conformational rotation in the presence of ATP, as it binds and releases from multiple binding sites along the actin filament. The actin filaments are comprised of actin monomers organized into an extended double helical chain. The flexible heads of the myosin molecules, projecting from the myosin filament, form the cross-bridges which attach and detach in a cyclical fashion at binding sites along the actin filaments.

(a)

Fig. 2.1 (a) Electron micrograph showing how skeletal muscle myofilaments (actin (thin) and myosin (thick)) are serially arranged into sarcomeres which span repeating Z-disks (very dark bands). The bold arrow shows the longitudinal axis of the fiber. Moving from a Z-disk, the banding pattern within a sarcomere corresponds to an I-band (no overlap region, thin filaments only), an A-band (overlap region of thick myosin filaments with actin filaments) and the intermediate dark M-band forming the middle of the thick myosin filament array. The hexagonal organization of myosin and actin filaments is seen in cross-section (following page) for (b) asynchronous and (c) synchronous insect flight muscle. A similar organization is observed for vertebrate and other invertebrate striated skeletal muscles. The insets in (b) and (c) show this hexagonal organization at higher power. Scale bars in (a) and (c) are 1 μm; the smaller scale bar in the inset in (c) is 0.1 μm. Abbreviations: L, lipid droplet; mc, mitochondrion; sr, sarcoplasmic reticulum; M, myofilaments; t, tracheole. ((a) Courtesy of Hans Hoppeler; (b) and (c) reproduced from Josephson et al. (2000), with permission of the Company of Biologists Ltd.)

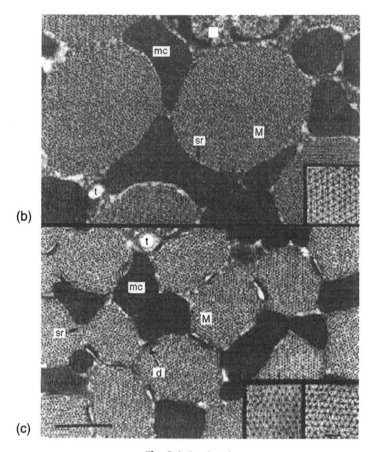

(b)

(c)

Fig. 2.1 Continued

Each cross-bridge cycle of myosin head attachment to actin and detachment involves the hydrolysis (splitting) of one ATP molecule. Chemical energy released from the ATP is converted into the force and rotational movement of the myosin cross-bridge head. As a result, myosin functions both as a machine for transforming chemical energy into mechanical work and an enzyme (myosin-ATPase) in the hydrolysis of ATP. Therefore rates of cross-bridge cycling (and ATP hydrolysis), which underlie the speed of muscle shortening and force development, can be qualitatively assayed by biochemical determination of the myosin-ATPase activity of a muscle's fibers. Details of the time-course of myosin binding to actin and their subsequent association with ATP, myosin head rotation and release from the actin binding site, and the force and stroke distance achieved by individual myosin cross-bridges remain under study, but it seems clear that ATP splitting is required for myosin cross-bridge detachment from the actin filament. In other words, the energy released by ATP hydrolysis occurs at the final step of the cycle when the myosin head detaches from actin and is then free to seek another binding site. ATP binding energizes the acto-myosin complex, enabling the subsequent conformational rotation of the myosin head, which is the basis of force development and shortening.

Because force development occurs only during rotational movement of the myosin head in one direction (toward the sarcomere midline), force development involves a relative sliding of the actin (thin) and myosin (thick) filaments with respect to each other (Fig. 2.2). Consequently, as force develops the sarcomere becomes shorter, resulting in an increased overlap between the thick and thin filaments. The multiple cross-bridge cycles of attachment, force generation and shortening, detachment and subsequent re-attachment are summed across the myofilament lattice of the sarcomere and along the length of the myofibrils when the muscle fiber is stimulated to contract, yielding an overall shortening of the muscle fiber and force generated at its ends. The amount of overlap between the myosin and actin filaments affects the number of cross-bridges that can be formed and therefore the amount of force that can be generated. The amount of force developed by a muscle also depends on the number of cross-bridges formed in parallel across the myofilament lattice. In practice, this is determined by the number of muscle fibers which are activated to contract by the nervous system(see Chapter 10). Owing to the highly conservative nature of actin and myosin in the skeletal muscles of diverse animals, the force that muscles can generate per unit area of activated fibers is quite uniform, typically being in the range 18–30 N/cm^2. Part of this range of force-generating capacity also reflects differences in the relative amounts of mitochondria, sarcoplasmic reticulum and other non-force-generating components which exist among different types of muscle cells.

2.1.2   Force–length relationship

Confirmation of the 'sliding filament model' for muscle contraction was based on X-ray diffraction studies of myofilament overlap of sarcomeres in relation to the amount of force developed by muscle fibers at different sarcomere lengths (Gordon *et al.* 1966). These studies revealed that increased overlap between actin and myosin (at shorter lengths) results in an increase in the active force that the muscle can develop ('ascending limb' of Fig. 2.2(a)) up to a maximal level. At this point force development reaches a plateau and then begins to fall at shorter lengths. The decrease in force at shorter lengths on the 'descending limb' of the force–length curve can be explained by excessive overlap, resulting in disruption in the myofilament lattice as the actin and myosin thick filaments increasingly interfere with one another, and as the myosin filaments push up against the Z-disks. The physical disruption of the regular spacing of their

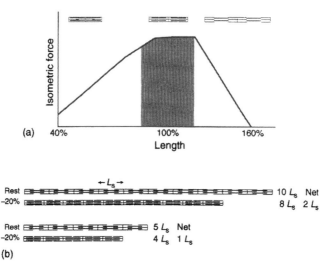

(a)

(b)

Fig. 2.2 (a) Isometric force–length curve for striated muscle, illustrating the plateau of maximum force at a length corresponding to maximum overlap between myosin and actin (Gordon *et al.* 1966). Force falls off at both long lengths, owing to reduced overlap, and at short lengths, owing to disruption of the myosin–actin spacing. (b) Schematic diagram showing how muscle fiber length and the number of sarcomeres in series affects overall muscle fiber shortening relative to the fractional shortening of its sarcomeres. In both cases, fractional shortening (shown below) is 20 per cent, but total shortening is twice as great for the upper fiber ($2\,L_s$ vs L_s) which is twice as long as the lower fiber.

hexagonal lattice prevents effective myosin cross-bridge binding to the actin filaments, causing a loss of force.

The force–length relationship depicted in Fig. 2.2(a) constitutes one of the fundamental properties of striated skeletal muscle. It suggests that muscle fibers (and, by implication, muscles) should be designed to operate at an intermediate range of sarcomere length (typically ±5 to ±15% of resting length; shaded region in Fig. 2.2(a)) which enables force development to remain near a maximum. Because there are more sarcomeres in series in longer muscle fibers than in shorter fibers, we can expect muscles which undergo greater length change in their function (i.e. to produce greater movement) to have longer fibers than muscles which function over shorter ranges of length. A longer fiber length allows a muscle to achieve a greater overall length change for a given fractional length change of its sarcomere (Fig. 2.2(b)). However, we shall see that the geometry of muscle attachment to the skeleton also greatly affects muscle shortening in relation to range of movement.

The linear dimensions of sarcomeres in vertebrate skeletal muscles are surprisingly uniform, typically falling in the range 2.0–2.8 µm. Consequently, functional adjustments for changes in operating length require changes in the fiber length of vertebrate muscles. Invertebrate skeletal muscles, on the other hand, exhibit a much greater diversity of sarcomere length (1.9–10 µm among arthropods and up to 40 µm in annelid worms (Hoyle 1983)). This means that differences in both sarcomere length and overall fiber length can affect the functional force–length relations of invertebrate muscles. Longer sarcomeres have longer myosin filaments, which means that more cross-bridges can be formed with neighboring actin filaments, so that longer sarcomeres are able to generate greater forces than shorter sarcomeres. Differences in sarcomere length also affect the speed of shortening of a fiber, with longer sacromeres generally contracting at slower speeds. The offsetting effects of force and speed mean that the amount of power that a muscle can produce per unit mass is independent of sarcomere length and is basically the same for all muscles.

In addition to their active force–length properties, muscles also possess passive elastic properties. When an inactive muscle is stretched from its resting length, it will resist the imposed stretch by developing force (Fig. 2.3(a)). Consequently, the force–length properties of whole muscles in the limbs of animals result from a combination of their *active* and *passive* components (Fig. 2.3(b)). While the active force–length properties of a muscle are critical to understanding the molecular basis of muscle force development and shortening, the passive elastic properties of muscles are also important to their behavior *in vivo* and cannot be overlooked. The passive elastic properties of muscles depends on their fiber architecture and the amount of connective tissue to which the fibers attach within the muscle. Muscles with short fibers and more extensive connective tissue (see discussion of pinnate versus parallel fibered muscles below) exhibit a steeper rise in their passive resistance to stretch compared with longer-fibered muscles which have less connective tissue investing them (Fig. 2.3(c)).

2.1.3   Force–velocity relationship

In addition to the effect of length, the velocity of fiber shortening (and lengthening) also affects the amount of force that a muscle can develop. With an increase in shortening velocity (right-hand side of curve in Fig. 2.4(a)), force falls off [in a hyperbolic fashion] from the level developed under isometric conditions (P_0 at zero shortening velocity) to (theoretical) zero at the muscle's maximum shortening velocity (V_{max}, normalized to 1.0 in Fig. 2.4(a)). The decrease in force at greater rates of fiber shortening probably results from a decreased ability of unbound myosin cross-bridge heads to bind successfully to actin sites as the speed of filament sliding increases, as well as the possibility that the force developed by each myosin head is also reduced.

We have seen above that the sliding filament mechanism of force generation suggests fiber shortening during force development. However, if a muscle

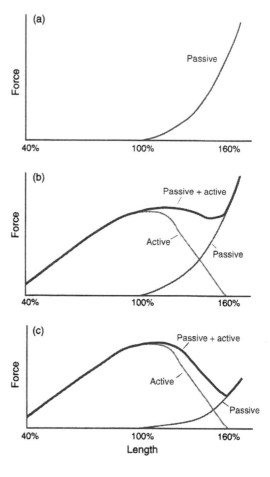

Fig. 2.3 (a) Passive force–length curve for a muscle. Resistance to stretch occurs as the muscle is lengthened from its resting length (100 per cent). The J-shaped curve is typical of compliant materials such as inactive muscle. (b) The active (similar to that shown in Fig. 2.2(a)) and passive force–length curves of a muscle combine to yield its overall force–length properties. This is typical of a pinnate muscle which has considerable connective tissue. (c) The combined active and passive force–length curves for a parallel-fibered muscle which has little connective tissue. Because of this, its passive component is much smaller than for a pinnate muscle (b) and there is a larger drop in tension at longer lengths.

contracts when loaded by a force which exceeds its isometric force, it will instead be actively lengthened. This can occur, for instance, when a person lands from a jump. Upon landing, the knee extensor muscles contract to prevent the legs from collapsing, but are lengthened as the knee initially flexes before re-extending to absorb the energy of the falling body. This occurs because the rate of body loading exceeds the development of isometric contractile force P_0 by the muscle fibers. Under these conditions, the force developed by the muscle rises sharply with an increase in the rate of lengthening. The additional force that the muscle generates while being actively lengthened is provided by a rapid stretch of the myosin cross-bridges attached to the actin filaments. This heightened level of force (up to 1.8 times peak isometric force) can only occur over very short distances of lengthening and short time periods. At greater distances of fiber lengthening, or longer time periods, the cross-bridges become

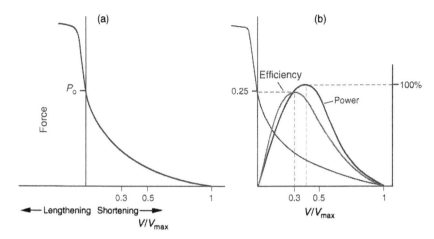

Fig. 2.4 (a) The Hill isotonic force–velocity curve for skeletal muscle. Note that shortening is defined as positive on the x-axis and that much greater forces can be developed than the isometric force P_0 when the muscle is actively stretched (but only over short distances). Velocity (and force) are often normalized to maximum shortening velocity V_{max} to allow comparison among different sizes and speeds of muscles. (b) The relationship of muscle efficiency (ratio of mechanical work output to metabolic energy input) and muscle power output (work per unit time) as a function of normalized muscle shortening velocity, with the force–velocity curve shown for comparison (light line). Muscles have zero efficiency and generate zero power when they contract isometrically (zero velocity) and at V_{max} (zero force). Whereas maximum efficiency is typically achieved at about $0.3V_{max}$, maximum power occurs at a slightly higher shortening velocity ($0.4V_{max}$). Thus muscle contraction rates are likely to vary according to whether an animal is maximizing endurance versus speed for a given activity.

detached and the additional elastic restoring force of the cross-bridge is lost. Indeed, excessive muscle stretch may be a leading contributor to muscle injury. Active muscle lengthening and the enhancement of force which it facilitates is believed to occur in the locomotion of many animals, but it normally happens over brief instances in time and short length changes.

The force–velocity relationship shown in Figure 2.4(a) represents a second basic property of skeletal muscle. The hyperbolic nature of this relationship was first described by A.V. Hill in 1938, after having previously won a Nobel Prize in 1922 for his work on the heat production of muscle. The force–length and force–velocity properties of muscle can reveal much about the functional organization and requirements of muscles in the limbs of an animal. However, it is important to bear in mind that these relationships are determined under well-defined but artificial experimental conditions compared with the dynamic conditions of locomotion. The force–velocity curve is determined from sets of measurements in which the muscle is stimulated and allowed to shorten against a fixed load over a very short range of length (<5 per cent). Such shortening contractions are referred to as being 'isotonic' (constant force). In living animals, muscles develop force and change length under dynamic conditions in which both force and length change at different rates during the period of a contraction. Consequently, the mechanical and physiological properties of muscles are likely to differ from the classical description of their force–length and force–velocity properties, which are based on idealized isometric and isotonic contractile states. Nevertheless, the properties which have emerged from these studies have proven exceedingly valuable for developing an understanding of the function and design of muscles in living animals. Considerably less is known about the dynamic properties of muscles, particularly when subjected to lengthening contractions.

2.1.4 Muscle power and efficiency

By plotting the product of force and velocity, the Hill force–velocity relationship indicates how muscle power varies as a function of isotonic shortening (Fig. 2.4(b)). This shows that, for most skeletal muscles, maximum power is developed at about $0.4V_{max}$. Thus muscles which function to generate mechanical power (for instance, the flight muscles of insects and birds and the axial muscles of fish, or the mantle muscles that power the jetting of squid) can be expected to operate with shortening velocities in this range if power output is to be optimized. The ratio of the work that a muscle performs to the chemical energy that it consumes (as ATP) during a contraction provides a measure of a muscle's efficiency. Because it is difficult to measure ATP use directly, extremely sensitive thermal measurements of the heat released by a contracting muscle (energy lost), relative to the amount of work performed, have provided a reliable alternative approach for measuring muscle efficiency (= work/ (work + heat)). These measurements have shown that the efficiency of skeletal muscle is maximal at a lower shortening velocity ($\sim 0.3V_{max}$) (Fig. 2.4(b)) than that observed which maximizes mechanical power. Consequently, if efficiency is to be maximized, which is important to an animal's economy of transport, its muscles can be expected to contract more slowly than if maximum power (for escape and speed) is required.

2.1.5 Muscle 'work loops': time-varying force–length behavior of muscles

While the isometric force–length and isotonic force–velocity relationships described above have proven extremely useful for understanding the design and properties of different muscles, neither provides a reasonable description of how muscles function under dynamic conditions. The natural movements underlying locomotor activity require that muscles be activated from a resting state to develop force and change length in a time-varying fashion. Neither the force that a muscle generates nor its length is likely to remain constant over the period of time when the muscle is active. The length changes of many muscles which propel the undulating body or reciprocating limbs of an animal can be approximated by a sinusoidal pattern of length change. The particular timing of muscle activation relative to its shortening and lengthening is critical to the net work that the muscle performs during a contraction cycle.

By plotting the force that the muscle develops relative to its change in length (Fig. 2.5(a)), the dynamic force–length behavior of a muscle can be observed. If the pattern of force–length behavior describes a counter-clockwise loop (Fig. 2.5(b)), the area within the loop is a measure of the net positive mechanical work that the muscle performs during each contraction cycle. If the pattern describes a clockwise loop, the area of the loop represents the net negative work performed by the muscle (Fig. 2.5(c)). In the former case, force development during muscle shortening (positive work) is greater than when the muscle is being lengthened (negative work shown by the hatched region in Fig. 2.5(b)). Consequently, the net work done by the muscle over a full contraction cycle is positive. In the latter case, force is greatest when the muscle is being lengthened. Hence greater work is done to stretch the muscle than when the muscle shortens to do positive work. As a result, the net work done *by the muscle* is considered to be negative (i.e. the muscle absorbs energy), and the term 'negative work' used. This is defined from the muscle's point of view—certainly not a definition that most physicists would like! For the ideal isotonic (constant force) conditions used experimentally to determine a muscle's force–velocity relationship (Fig. 2.4(a)), the muscle's time-varying force–length behavior would look like the graph shown in Fig. 2.5(d). However, as noted above, instantaneous force development followed by constant force during shortening and then instantaneous relaxation are unrealistic for muscles operating under natural locomotor conditions.

The force–length patterns described by the behavior of muscles which develop force under a sinusoidal or similar oscillating patterns of length change represent muscle contraction cycles termed 'work loops'. Studies of muscle work-loop performance have recently proven extremely valuable for studying the properties of muscles under the dynamics of muscle length change and force development believed to operate under *in vivo* conditions of locomotor activity. Muscles that function to do work and generate power (equal to the product of work per contraction and the contraction frequency), such as the axial body muscles that power fish swimming, the leg muscles that power locust and frog jumping, and the flight muscles of birds, characteristically have broad counter-clockwise work loops similar to that shown in Fig. 2.4(b). However, muscles which operate as brakes to absorb energy, such as when overcoming the inertia of a moving limb or decelerating the body's motion,

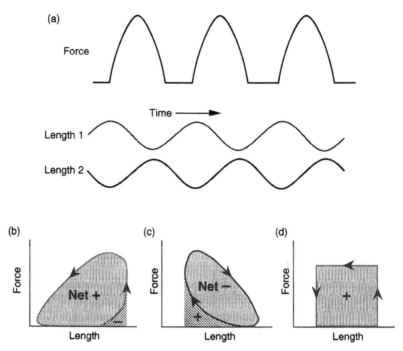

Fig. 2.5 (a) Time-varying patterns of muscle force and length change that are common to the *in vivo* locomotor behavior of many muscles, particularly those which undergo oscillatory length change and perform mechanical work. Note that these patterns are not well represented by the classical force–length (Fig. 2.2) and force–velocity (Fig. 2.4) properties of skeletal muscle. (b) Force versus length behavior of a muscle undergoing length oscillation (Length 1) relative to the phase of force output shown above. This describes a counter-clockwise 'work loop'. The shaded area in the center of the loop is the net work that the muscle performs over the contraction cycle. The hatched area at the lower right is the negative work (energy absorbed) by the muscle while being lengthened during this phase of the cycle. The negative work is subsequently recovered when the muscle shortens, yielding the net work within the loop. (c) Clockwise 'negative work loop' formed by the Length 2 oscillation pattern above. This results from a delay in the phase of muscle shortening compared with the Length 2 pattern. This causes the muscle to do little work while it is shortening (hatched region: force is low or zero) and absorb energy by developing greater force while being lengthened. (d) This represents an idealized isotonic shortening work profile for a muscle that can instantaneously develop force, shorten at a constant force and then relax instantaneously before being lengthened. The work loop in (b) is a far more realistic representation of the actual behavior of locomotor muscles performing positive work.

can be expected to generate clockwise work loops (Fig. 2.4(c)). Finally, as we shall see in Chapter 3, muscles may also function to contract economically by developing force with little or no change in length, so that they do little net work. When this occurs, the force–length behavior of a muscle results in little or no area contained within the loop. Indeed, in this case the term 'work loop' is

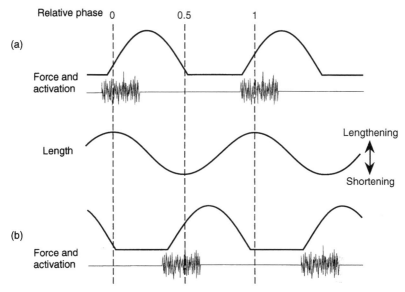

Fig. 2.6 Timing or phase of muscle force and (EMG) activation relative to length change. (a) An activation pattern characteristic of a muscle which shortens while developing force to perform positive work (counter-clockwise work loop). (b) A pattern characteristic of a muscle which absorbs energy by developing its force while being lengthened (negative clockwise work loop).

a misnomer, and the overall pattern of force–length behavior is actually much more similar to that of a spring (see Fig. 1.1(a)).

The timing and duration of muscle activation relative to its length change are also key to a muscle's work performance. Most commonly, the timing of muscle activation is defined in terms of its phase relative to when the muscle begins to shorten (phase 0). Activation of a muscle immediately in advance of its shortening (negative phase: −0.1 to 0) and with a duration lasting midway through shortening (relative phase: 0.25) is most favorable for effective force development and subsequent work during shortening (Fig. 2.6(a)). This is because the initial period of muscle activation occurs while the muscle is briefly being stretched or develops force under near isometric conditions. From the force–velocity relationship (Fig. 2.4(a)), we know that muscles develop greater force under these conditions. If activation of the muscle is delayed until after it begins to shorten (relative phase: 0 to 0.5), its ability to develop force and do work during shortening will probably be reduced. If activation of the muscle occurs later, when it nears the end of shortening, and lasts into a substantial period of lengthening, the muscle's force–length behavior will result in a clockwise (negative) work loop (Fig. 2.6(b)). Finally, if a muscle's activation is limited to when it undergoes little or no length change, it will develop force isometrically and do little or no work.

Consequently, it is important to bear in mind that not only are the physiological properties and fiber architecture (see section 2.1.7) of a muscle important to its contractile performance, but the timing and duration of its activation will greatly influence the nature of work that the muscle does. This also has important implications for how the nervous systems of animals control their locomotor movements.

2.1.6 Muscle excitation–contraction coupling

Except for the asynchronous flight muscles of insects (see Chapter 4), most skeletal muscles are activated directly by motor nerves which innervate a subpopulation of fibers within a muscle (Fig. 2.7(a)). In vertebrates, depolarization of the muscle fibers by action potentials transmitted by the motor nerve across its motor endplates leads to release of calcium ions (Ca^{2+}) from the sarcoplasmic reticulum (SR), which extends throughout the muscle cell (Fig. 2.7(b)). Inward flow of the cell membrane's depolarization is conducted via a transverse-tubule system to deeper regions of the SR, ensuring that myofilaments at the fiber's center are activated together with those closer to its surface. Vertebrate skeletal muscle fibers themselves respond in an all-or-nothing fashion as the spread of depolarization of the fiber spreads rapidly along the fiber's length from the motor endplate. As a result, they are considered to be 'twitch-type' fibers. The firing rate of action potentials transmitted to the motor endplates influences the magnitude and time course of force which the muscle develops by regulating the amount of calcium that is released from the SR. Nearly all vertebrate skeletal muscles that power locomotion are comprised of 'twitch' fibers, which develop a characteristic pattern of force in response to endplate depolarization (Fig. 2.8(a)). Multiple stimuli at slow to moderate frequencies result in an 'unfused tetanus' which yields an elevated, but rippled, force pattern (Fig. 2.8(b)). With an increase in stimulation frequency, the magnitude of force increases further (a property termed 'summation') and a smooth force plateau, or 'fused tetanus', is achieved (Fig. 2.8(c)). The stimulation frequency required to elicit a fused tetanus is greater in faster contracting muscle fibers. Faster contracting muscles develop force more rapidly than slower contracting muscles, but fatigue more quickly. In most instances, vertebrate skeletal muscles are believed to generate fused tetanic contractions during locomotion.

In the case of invertebrate muscles, the majority of the fibers do not usually respond in an all-or-nothing fashion. Instead, they respond to nerve stimulation by developing graded patterns of depolarization produced by multiple motor nerve terminals which are distributed along the length of the muscle fiber. Local motor junction potentials sum over the length of the fiber to produce a net depolarization in the fiber as a whole. In addition, invertebrate skeletal muscle fibers may also receive inhibitory input from their motor nerves, in contrast with vertebrates whose motor nerves act solely in an excitatory fashion. Therefore the graded activation of invertebrate muscle fibers provides a fundamentally different mechanism for controlling the force output

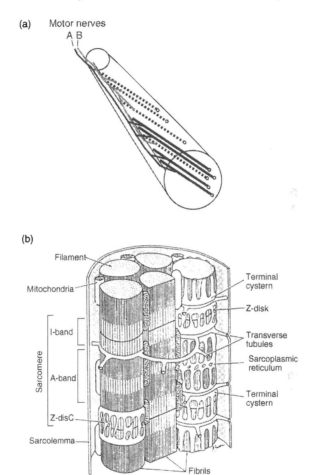

Fig. 2.7 (a) Schematic drawing showing two vertebrate motor units formed by the motor nerves (A and B) and the subpopulation of fibers within the muscle that each nerve innervates separately. (b) Diagram showing the SR in relation to the sarcolemma (muscle cell surface), the myofibrils and the transverse (T) tubules, which conduct the depolarization of motor activation inwardly to activate all myofibrils. (c) Schematic drawing showing a vertebrate motor endplate which releases acetylcholine (Ach) as the excitatory neurotransmitter. Ach release causes an all-or-nothing depolarization of the muscle cell which is propagated rapidly along the fiber's length. The T tubules conduct this charge inwardly to elicit Ca^{2+} release from the SR to trigger cross-bridge cycling and muscle contraction. ((b) and (following page, C) reproduced from Loeb and Gans (1986), with permission from University of Chicago Press.)

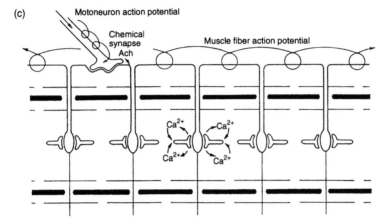

Fig. 2.7 Continued

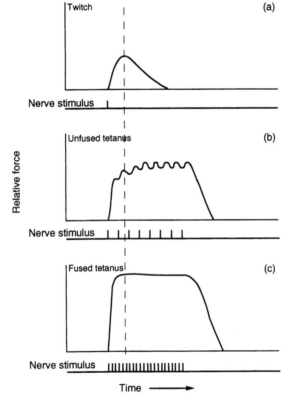

Fig. 2.8 Muscle force response to varying frequencies of stimulation: (a) a single twitch stimulus; (b) unfused tetanus; (c) fused tetanus. These patterns are for twitch-type muscle fibers. See text for details.

of a muscle. Nevertheless, the underlying basis for force development is the same in both vertebrates and invertebrates, depending on the release of Ca^{2+} from the SR in response to depolarization. Release of Ca^{2+} is voltage dependent. Consequently, the greater the depolarization of the fiber, the more Ca^{2+} is released and the faster and greater is the force that the muscle develops.

The release of Ca^{2+} from the SR (Fig. 2.1(a)) catalyzes myosin's hydrolysis of ATP in the cross-bridge cycle by means of Ca^{2+} binding to troponin C, a protein associated with actin which exposes the myosin binding site on the actin filament when complexed with Ca^{2+}. Therefore cross-bridge cycling is enabled by SR release of Ca^{2+} which switches on the development of force and muscle shortening via myosin hydrolysis of ATP. Muscle relaxation occurs by active ATPase-dependent pumps in the membrane of the SR which remove Ca^{2+} from the cytoplasmic myofilament lattice and re-sequester it into the SR. Active Ca^{2+} uptake by these pumps is an ongoing process, so that maintenance of force requires ongoing depolarization by the motor nerve which stimulates Ca^{2+} release from the SR in order to maintain an elevated concentration of cytosolic Ca^{2+}. In general, for most muscles the energy cost of Ca^{2+} release and uptake by the SR is a small (5–20 per cent) but significant fraction of the energy cost associated with force generation. Interestingly, repetitive activation of a muscle does not appear to incur a proportional Ca^{2+} pumping cost. The basis for this is still not well understood, but it means that most of the energy cost associated with repetitive muscle contraction during locomotion is tied to the cost of force generation associated with cross-bridge cycling.

2.1.7 Motor units and muscle fiber types

The motor nerve and the set of muscle fibers that it innervates comprise a **motor unit** (Fig. 2.7(a)). In nearly all mammalian, avian and reptilian muscles, adult muscle fibers are innervated by a single motor nerve; however, in some fish and amphibian muscles, as well as those of various invertebrates, a muscle fiber may be innervated by more than one motor nerve (termed **polyneuronal innervation**). In those animals lacking polyneuronal innervation, the fibers innervated by the motor nerve (i.e. those comprising the motor unit) share similar contractile properties. This is believed to result, in part, from a trophic influence of the motor nerve on the fibers. Therefore modulation of force output by twitch-type skeletal muscles is achieved by the recruitment of motor units by the nervous system within a muscle (covered in detail in Chapter 10). The number of fibers innervated by a motor nerve can vary, with finer control of force being achieved by having a larger number of small motor units. For much smaller invertebrate muscles, such as the steering muscles of flying insects, which may be innervated by a single motor nerve, the number and frequency of nerve stimuli, as well as the time course of depolarization, are varied to modulate force.

Invertebrate and vertebrate skeletal muscle fibers can generally be classified as either twitch or tonic. Twitch fibers, as noted above, are those with which we are most familiar. When activated they develop force, and when their stimulation ceases they relax. Hence they contract in a phasic manner in response to neural stimulation. In contrast, tonic muscle fibers which are slow can maintain their tension for long periods of time, well after their neural activation has ended. As a result, slow tonic muscle fibers provide a means for adjusting and maintaining tension at low energy cost for long time periods. Slow tonic fibers are found in muscles of invertebrates, as well as fish, amphibians, reptiles and birds, but are believed to be absent in mammals. In addition to having twitch fibers similar to vertebrates, invertebrate muscles mainly contain non-twitch fibers which can also be distinguished as slow versus fast contracting. These are fibers which can respond to motor stimulation slowly or more rapidly, through graded depolarizations of their membrane. The fibers used in locomotor movements have contraction rates that are much faster than the slow tonic fibers used for posture maintenance. Here, we shall mainly be concerned with the properties of twitch-type skeletal muscle fibers, and the faster contracting non-twitch fibers of invertebrates, in relation to how they are recruited for adjustments in muscle force, speed and endurance.

Both vertebrate and invertebrate skeletal muscles commonly have twitch fiber populations which possess differing contractile and metabolic characteristics. Distinguishing among these features has led to a series of descriptive classifications of muscle fiber types. Although these classification schemes suggest discrete types of muscle fibers, in fact considerable variation can exist within a population of fibers of a given type. Consequently, differences between fiber types are often qualitative, yielding more of a continuum of properties across the spectrum of muscle fiber types. Nevertheless, differences in motor recruitment and function can usefully be ascribed to differences in fiber type within and between muscles within a species and can often have profound physiological and ecological significance when comparing species. The characteristics and properties of twitch muscle fiber types have been most thoroughly studied in mammals. Therefore the classification scheme used here is largely based on these studies. Although this reflects the bias of twitch mammalian fiber types, it provides a conceptual scheme which can usefully be extended to the broader diversity of skeletal muscle fibers found in other animal groups. Characteristics of non-twitch invertebrate muscle fibers will be discussed in Chapter 10 when we consider the neural control of motor function.

Three main types of twitch muscle fiber are generally recognized within mammals, birds and reptiles: slow-oxidative (SO), fast-oxidative glycolytic (FOG) and fast-glycolytic (FG) (Fig. 2.9). Table 2.1 lists several key contractile, metabolic and cytological features of these three types of fibers. Recently, more fine-grained resolution of muscle fiber types has been established using protein

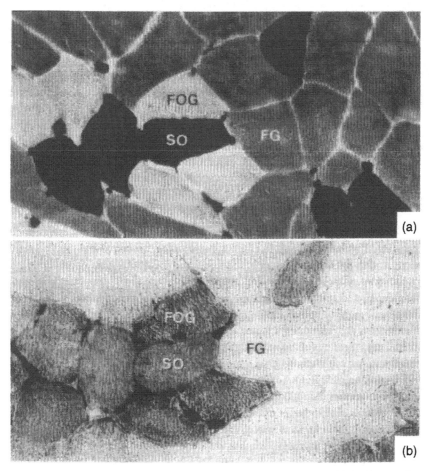

Fig. 2.9 Comparison of histochemical fiber patterns in (a) the superficial and (b) the deep portions of cat semitendinosus muscle, showing staining characteristics after myosin ATPase reaction at pH 4.1. Under acidic conditions, SO fibers stain darkly, FG fibers are intermediate and FOG fibers are light. See text for details. (Reproduced from Hoppeler *et al.* (1981), with permission from Elsevier/North-Holland Biomedical Press.)

biochemistry to characterize the different types of myosins (termed myosin **isoforms**) which are expressed within different fibers. Using these results, SO fibers are also often referred to as type I fibers, FOG fibers as type IIa (also IIx) fibers and FG fibers as type IIb fibers. Once again, these classifications represent *qualitative* descriptions of the contractile and metabolic properties of the fibers. Myosin which stains darkly under basic conditions of incubation correlates with myosin which has a fast ATPase rate (Fig. 2.9) and hence is characteristic of fibers which contract rapidly (i.e. fast cross-bridge cycling rate).

Table 2.1 Twitch (phasic) skeletal muscle fiber types (based largely on mammalian patterns)

Property/feature	Slow-oxidative (SO) (Slow twitch type I)	Fast-oxidative glycolytic (FOG) (Fast twitch red type IIa)	Fast-glycolytic (FG) (Fast twitch white type IIb)
Contractile			
Force/fiber	Low	Intermediate	High
Speed of shortening and force development	Slow	Moderate to fast	Fast
Endurance to fatigue	High	Intermediate	Low
Fiber diameter and force/fiber	Small	Intermediate to large	Large
Metabolic			
Myosin ATPase (acid pH)	Low	High	High
Myosin isoform (heavy chain)	I (identical to cardiac β)	IIa and/or IIx	IIb
Citrate synthase and SDH (Krebs cycle)	High	Intermediate	Low
GDP	Low	Intermediate	High
Myoglobin	High	High	Low
Muscle			
Glycogen	Sparse	Intermediate	Many glycogen granules
Number of mitochondria and capillaries	Many	Intermediate	Few
SR	Simple	Intermediate	Well developed

SDH, succinate dehydrogenase; GDP, glycerol diphosphatase; SR, sarcoplasmic reticulum.

More oxidative fibers have slower myosin ATPase rates, which means that they develop force and shorten more slowly. A slower rate of shortening is also associated with their reliance on aerobic synthesis of ATP via mitochondrial oxidative phosphorylation. Hence, SO fibers show a high quantity of oxidative enzymes, such as succinate dehydrogenase (Fig. 2.9) or citrate synthase. In contrast, FG fibers store large amounts of glycogen and have a high quantity of glycolytic enzymes, such as glycerol diphosphatase, enabling them to generate ATP rapidly via anaerobic metabolism. However, because of their metabolic differences FG fibers fatigue quickly (due to metabolic acidosis), whereas SO fibers can contract for long periods without becoming fatigued. As their name suggests, FOG fibers generally possess intermediate properties.

Associated with their metabolic reliance on aerobic ATP synthesis, SO fibers are also more richly supplied with capillaries and myoglobin (reflecting a high capacity for oxygen delivery and transport) and contain many more mitochondria than FG fibers. FG fibers, on the other hand, have a more extensive well-developed SR, associated with the need for rapid Ca^{2+} activation and uptake for fast contraction speed. Recruitment among fiber populations with these differing metabolic and contractile properties allows animals to use particular sets of muscles or muscle fibers for particular tasks (see Chapter 10 for a more detailed discussion). Rapid burst activity, as during escape or prey capture, is best powered by FG fibers, whereas slower movements and control of posture are better suited to recruitment of SO fibers. Finally, it is worth emphasizing again that the slow-twitch oxidative fibers described in this way are quite distinct from and not to be confused with the slow-*tonic* fibers found in certain muscles of invertebrates and vertebrates.

Important taxonomic differences also exist in terms of fiber types. As ectotherms, most reptilian species rely on burst activity rather than sustained locomotion in order to obtain food and avoid predation. Not surprisingly, the leg muscles of lizards that have been investigated have a predominance of FG fibers that are well suited to burst anaerobic activity. On the other hand, mammals and birds are able to sustain locomotor activity that is fueled by aerobic metabolism for much longer periods. Consequently, their muscles have much higher proportions of oxidative (SO and FOG) fibers. Most spectacularly, the high wing beat frequency of hummingbirds (45–60 Hz) requires that nearly 50 per cent of the muscle volume is occupied by mitrochondria (the other half is comprised of myofilaments and SR) and that nearly all of their flight muscle is comprised of FOG fibers. This is also true of many small to moderate-sized birds. Even in moderate-sized birds, such as pigeons, up to 85 per cent of the flight muscle is represented by FOG fibers. Clear differences exist among mammals as well. Whereas canids (dogs, coyotes, wolves) have high proportions of oxidative fibers allowing them to hunt and forage actively for long periods, felids (cats, lions) have relatively few oxidative fibers, relying on stealth and burst pursuit to catch their prey. Such differences clearly distinguish the temperaments and personalities of our most popular pets!

2.1.8   Muscle architecture

In addition to a muscle's intrinsic force–length and force–velocity properties and its fiber type characteristics, its fiber architecture is also important to its contractile function. Muscles are traditionally divided into two basic classes based on their architecture: parallel-fibered and pinnate-fibered muscles (Fig. 2.10). Parallel-fibered muscles have fibers which run end-to-end within the muscle in a direction parallel to its axis of force transmission. Typically, parallel-fibered muscles have relatively long fibers for their size and attach directly to the skeleton with little or no external tendon. In contrast,

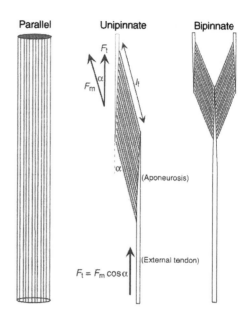

Fig. 2.10 Comparison of muscle architectures. In contrast with parallel-fibered muscle, the fibers of pinnate muscles exert force at an angle to the muscle and tendon long axis. The effective force is a product of the cosine of the pinnation angle α of the muscle fibers. F_t, tendon force; F_m, muscle fiber force; l_f, muscle fiber length. See text for details.

pinnate-fibered muscles are comprised of shorter fibers which run at an angle to the muscle's principal axis of force transmission. Pinnate muscles also often attach to the animal's skeleton via an external tendon. They can have quite complex architecture. Uni-pinnate muscles have fibers oriented at a similar angle in a single plane which attach to the distal tendon. Bi-pinnate muscles have two sets of fibers which angle at (roughly) mirror images to each other, with each set attaching to a central distal tendon. Multi-pinnate muscles have more complex fiber architectures with varying planes of angled fibers. In general, muscles that are more pinnate also have shorter fibers.

Owing to differences in fiber length, parallel-fibered muscles of equal mass (and volume) have a smaller fiber cross-sectional area compared with pinnate muscles. Because all striated muscles depend on the same myofilament interaction for force generation, the force developed by a muscle generally varies in direct proportion to its fiber cross-sectional area. In other words, the maximum isometric stress (force/fiber area) which striated muscles can develop is largely constant across a broad diversity of invertebrate and vertebrate animals. Consequently, because of their shorter-fibered architecture, pinnate muscles of equal mass are capable of generating greater forces than parallel-fibered muscles. Although pinnate muscles suffer some loss of force transmission because of the angle of their fibers, this is more than offset by their greater fiber area.

The cross-sectional area A_m of a parallel-fibered muscle can be determined from its mass m and mean fiber length l_f, provided that the density is known

(1060 kg/m^3 is a typical value for striated muscles):

$$A_m = m/\rho_m l_f. \tag{2.1}$$

The effective 'physiological' cross-sectional area $^*A_m$ of a pinnate muscle can be calculated from the pinnation angle α of the muscle fibers (Fig. 2.10):

$$^*A_m = A_m \cos \alpha. \tag{2.2}$$

For example, a parallel-fibered muscle with fibers 10 times as long as those of a pinnate muscle of equal mass and a pinnation angle of 20° will have a fiber area that is only 1/9.4 (= 1/(10 cos 20°)) or 0.106 times that of the pinnate muscle. Correspondingly, the effective force F_t that the pinnate muscle transmits to its tendon equals the force F_m developed by the muscle's fibers multiplied by cos α, which in this case is 9.4 times greater than the parallel-fibered muscle. However, an important advantage of a parallel-fiber architecture is that it provides a greater range of shortening (in the example above, it would be 10 times greater than that provided by the pinnate muscle). In terms of muscle work, the increase in force achieved by a pinnate architecture is countered by the decreased shortening capacity of its fibers. Consequently, all muscles have generally the *same* capacity for doing work on a per unit mass basis, irrespective of their fiber architecture.

2.1.9   Effect of muscle architecture on active muscle volume and energy use

The fiber architecture of a muscle also affects its rate of energy use based on the volume of muscle that must be activated to generate a given force. Long parallel-fibered muscles require more energy to generate a given force than short pinnate-fibered muscles. This is because longer fibers require the formation of more cross-bridges to transmit force along their length than short fibers. Consequently, the ATP cost per force generated, a measure of a muscle's **force economy**, is greater the longer are a muscle's fibers. For example, two muscles of equal volume but with fiber lengths that differ by a factor of 3 will also have fiber cross-sectional areas A_f which differ by a factor of 3 (Fig. 2.11). To generate a given force, both muscles must recruit the same A_f; however, this will require the formation of three times the volume and three times the number of cross-bridges in the long-fibered muscle. Consequently, given that both muscles contract under similar conditions, the cost to gener- ate a given force will be three times as great in the long-fibered muscle. As we shall see, this simple effect of muscle geometry on ATP utilization can have profound effects on the energy use of a muscle and its role in locomotor function.

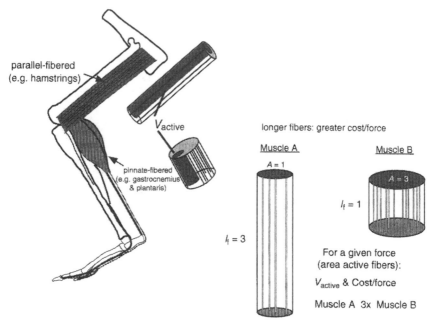

Fig. 2.11 A comparison of how fiber length affects the volume of muscle that must be recruited to generate a given force. For a given force, an equivalent number of fibers (equal fiber area) must be recruited. This means that in muscle (a), which has fibers three times as long as those in the shorter more pinnate muscle (b), three times the volume of muscle must be activated, which incurs three times the energy cost per unit force. Thus muscles with shorter fibers can achieve greater force economy. See text for details.

2.2 Skeletons

Locomotor movement requires that internal muscle forces are effectively transmitted to the external environment. This is achieved by muscles attaching to a rigid or incompressible skeleton. In both unicellular and multicellular animals, this is accomplished by having a combination of rigid compression-resistant and flexible tension-resistant elements. Multicellular animals possess three types of skeletons: endoskeletons, exoskeletons and hydrostatic skeletons. Of these, our own vertebrate endoskeleton is the most familiar. Endoskeletons and exoskeletons (characterized by the cuticle of arthropods and the shells of bivalve molluscs) have material properties which enable them to transmit force with very little deformation. The hydrostatic skeletons of many invertebrates—components of which are also found in vertebrates—use water or the fluid contained in an internal body cavity (coelom) as the compressive component. Hydrostatic compressive support, against which arrays of actin filaments and microtubules can act to transmit tensile forces, is also important to single-celled organisms. Because of their rigidity, compressive skeletal elements articulate at joints which allow and proscribe a certain range of motion based on the shape of the joint. In vertebrate endoskeletons, a specialized hydrostatic compressive-resistant tissue, termed **cartilage**, has evolved to provide low-friction movement and effective load transfer across the ends of articulating bones. Tensile elements involve collagen- and chitin-based ligaments and tendons which help to stabilize the joints between skeletal elements and link muscles to the skeleton.

The design of animal skeletons is linked to their role in providing effective force transmission for locomotion and body support. Consequently, they are best considered in relation to the muscles which power locomotion in multicellular animals. As with many other features of an animal's function, size plays a major role in the design and function of the skeleton. Hence, both the material properties and structural form of skeletons are important to their organization and function.

2.2.1 Vertebrate endoskeletons

Internal bony elements comprise the skeletons of vertebrates. Except for certain elements of the skull, the skeleton first develops and grows as cartilaginous elements which later ossify by means of calcification of the cartilage and subsequent mineralization of collagen (a process known as 'endochondral ossification'). The bones of mature animals are generally two-thirds mineral (hydroxyapatite) and one-third type I collagen. The mineral phase of bone is a hydrated calcium phosphate salt secreted by bone cells which crystallizes and grows on the type I collagen fibers that these cells also deposit into the extracellular matrix. As the collagen becomes mineralized, the bone's stiffness (ratio of load to deformation; see Chapter 1) increases.

Like many biological materials, bone is a 'composite' material. Biological composites are comprised of two or more components (in this case collagen fibers and mineral). The mechanical properties of such materials can be dramatically altered by varying the relative amounts of the two components and by varying how they are organized at a macroscopic level. Further mineralization of bone would increase its stiffness and probably make it stronger, but this would also diminish its ability to absorb the energy of impact loads (making it more brittle). Mineralization levels for a variety of limb bones have been found to be maintained within a fairly narrow range (63–70 per cent), suggesting that this trade-off between stiffness, strength and energy absorption is important (Fig. 2.12(a)). Consequently, bone is a rigid supportive material providing stiffness to transmit forces effectively, yet it is also capable of deforming sufficiently (<1 per cent) to allow it to absorb a substantial amount of energy before failing (Fig. 2.12(b)). In contrast, highly brittle materials like glass are often strong but easily break (i.e. have a high failure stress but absorb little

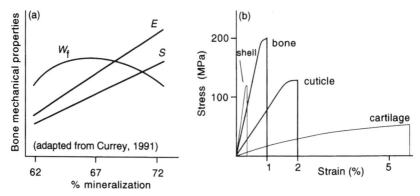

Fig. 2.12 (a) Comparison of bone properties (*S*, strength or maximum stress; *E*, elastic modulus or stiffness; W_f, work of fracture) as a function of bone mineralization. (b) Comparison of the stress–strain behavior of various skeletal tissues. See text for details. ((a) modified from Currey (1984).)

energy owing to their extremely high modulus; see Fig. 1.2(c). More compliant materials, such as cartilage, absorb energy by undergoing considerable deformation (Fig. 2.12(b)), but therefore are not rigid enough to function effectively for force transmission as a lever or over long distances. Consequently, bone appears to have evolved a level of mineralization which enables it to meet these competing requirements. On the other hand, as a relatively thin tissue layer at the articular end of a bone, cartilage functions well to absorb energy and distribute loads transmitted across a joint. The cartilage found in synovial joints (which represent most of the mobile joints of the vertebrate body) also provides excellent lubrication through the synovial fluid that is secreted into the joint cavity, taken up by the cartilage and squeezed into a film between the load-bearing surfaces of the articular ends of the bones.

While bone fracture does occur, it is a relatively rare event. Failure of important skeletal elements probably poses a significant biological cost to an animal's fitness. This cost is successfully avoided by the evolution of material properties and a structural design sufficient for a skeleton to bear the loads normally applied during an animal's lifetime. As a result, natural selection has favored skeletons which operate with safety margins endowing them with strengths that exceed their maximum likely loads by as much as a factor of 3–5, or even more. In addition to the large size range of growth that they allow, a distinct advantage of bony endoskeletons is their ability to be repaired, so that fracture need not incur a permanent and probably fatal cost.

2.2.2   Invertebrate exoskeletons

The muscles and apodemes (chitin-like tendons) of invertebrates which possess an exoskeleton necessarily attach to the inside of the skeleton. By containing the soft tissues inside, exoskeletons have the advantage of providing good protection and, in the case of insects, resistance to desiccation. On the other hand, they severely limit the animal's growth and are susceptible to damage on their exterior surface. Further, animals with exoskeletons have no means of repairing damage once it has occurred, other than by discarding their skeleton altogether. The rigid cuticle of arthropod exoskeletons must be shed at regular intervals (molt) to allow for growth. The calcium carbonate shells of bivalve molluscs and brachipods, on the other hand, grow by accretion at the border of the shell. Growth in size of the animal's soft parts is facilitated by the shells having geometries (cone, hemi-ellipsoid) which allow an increase in size while maintaining a similar shape (isometry).

The cuticle of arthropod exoskeletons, like bone, is also a composite material. Nearly all arthropod exoskeletons contain approximately 15–20 per cent stiff polysaccharide chitin fibers (similar in structure to cellulose) embedded in a protein matrix. In crustaceans, additional stiffness is achieved by the incorporation of calcium carbonate. However, the cuticle of most exoskeletons is stiffened mainly by 'tanning' (sclerotization), a process in which the cuticle matrix proteins become cross-linked by quinones and dehydrated to provide stiff reinforcement of the chitin fibers. The chitin fibers themselves are bonded to each other in cross-ply layers, much like plywood. (In apodemes, the chitin fibers lie parallel to each other, similar to the organization of collagen fibers in vertebrate tendons.) Tanned cuticle has approximately half the stiffness and strength of bone, but as a result absorbs more energy before failing (Fig. 2.12(b) and Table 2.2). Joints within the exoskeleton are achieved by hinged articulations between adjacent rigid cuticle elements with flexible (untanned) cuticle lying over the joint to allow movement.

Owing to the absence of a significant organic component, the calcium carbonate shells of bivalve molluscs are extremely stiff but fairly brittle. This probably reflects the continual need to resist damage by boring organisms

Table 2.2 Mechanical properties of some skeletal materials

	Tensile strength (MPa)	Stiffness (GPa)	Toughness (kJ/m^2)	Failure strain (%)	Density (kg/m^3)
Mammalian compact bone	150–180	18–20	2.0	0.9	2000
Locust leg cuticle	95	9.5	1.0[a]	1.0	12
Locust apodeme	600	19	6.3[a]	3.2	
Crab carapace	32	13	0.5[a]	0.3	1900
Mollusc shell	30–100	30–80	0.6	0.1	2700

Adapted from Wainwright *et al.* (1976).

common to their aquatic environment and the relative unimportance of resisting impact loads (Fig. 2.12(b)). The former capacity is favored by having an extremely hard skeletal surface, which the calcium carbonate provides. When the shells are used for locomotion, as in scallops, the shells articulate by a simple hinge joint at their base allowing them to be clapped together to expel water from the animal's mantle cavity (see Chapter 4).

2.2.3 Hydrostatic skeletons

Hydrostatic skeletons are one of the most widely distributed types of skeleton in the animal kingdom (being especially prevalent in coelenterates, worms, molluscs) and also form important components within the endo- and exoskeletons of other animals. For example, the articular disks present between the vertebrae of vertebrate animals represent hydrostatic components that are effective in resisting compressive loads transmitted between the vertebrae. Most commonly, a hydrostatic skeleton is formed by a pressurized fluid-filled cavity surrounded by a tension-resisting fiber-reinforced skin or wall. In the case of vertebrate articular disks, the annulus of the disk has collagen fibers arranged in layers that resist the gel-like nucleus pulposus contained inside. The hydrostatic skeletons of coelenterates, worms and other animals which must rely on their skeletons for movement as well as support are commonly associated with two sets of antagonistic muscles (e.g. circular and longitudinal) found, together with the fiber-reinforced layers, within the wall of the skeleton.

By being incompressible, the fluid contained inside transmits the forces developed by the animal's muscles to enable both changes of body shape and locomotor movement. Consequently, animals with hydrostatic skeletons do not depend on the rigid mechanics of force transmission that is characteristic of animals with endo- and exoskeletons. In most instances, locomotion is slow and non-dramatic. However, high-performance jetting locomotion has been evolved within certain cephalopod molluscs (squids) which use hydrostatic pressure developed within the mantle cavity to move at high speeds (see Chapter 4).

2.2.4 Skeletons as jointed lever systems

In animals with rigid skeletons, muscles produce movement and transmit force by developing moments at specific joints. Because muscles can only generate tensile (pulling) forces, reciprocal movements of appendages (e.g. flexion and extension of a joint) require the action of opposing, or **antagonistic**, sets of muscles (Fig. 2.13(a)). As noted above, this is also true for animals with hydrostatic skeletons. In some animals, a single muscle may power locomotion, but this muscle is always associated with a passive elastic element (in the case of scallops, an abductin pad) which acts as the muscle's antagonist by storing elastic strain energy when the muscle contracts and restoring this energy when the muscle relaxes to produce the reciprocal motion of the skeleton.

The joint moment (or torque) developed by a muscle (Fig. 2.13(b)) depends on the force F and the mechanical advantage or moment arm r, measured as the perpendicular distance from the vector of force to the axis of joint rotation:

$$M = F \times r. \tag{2.3}$$

Correspondingly, muscle shortening ds can be related to joint displacement (Fig. 2.13(d)) by the relationship

$$ds = rd\theta \tag{2.4}$$

or, for velocity,

$$ds/dt = rd\theta/dt \tag{2.5}$$

where dθ is the change in joint angle measured in radians (1 radian $= 180/\pi$). Equations (2.4) and (2.5) are approximations, but hold well for small θ (<0.5 rad). Attachment further from the joint's center of rotation increases a muscle's mechanical advantage but also increases the distance that it must shorten to produce a given angular displacement of the joint. Consequently, *the moment which a muscle can develop varies inversely with the range and speed of joint motion that it can produce.* Therefore the trade-off in force versus speed of movement exists at two levels of muscle function:

- the muscle's intrinsic force–velocity relationship (section 2.1.3)
- the force–velocity relationship resulting from the lever mechanics of joints.

Whereas the mechanical advantage of the limb shown in Fig. 2.13(b) is twice that of the limb in Fig. 2.13(c), allowing it to develop twice the force G against the ground for a given extensor muscle force, the range and speed of motion of the foot is 50 per cent less (compare Figs 2.13(d) and 2.13(e)). The length of the skeletal segment to which a muscle or its tendon attaches also affects its relative mechanical and displacement advantage for generating force and movement at the end of the segment. The longer the distal skeletal segment (compare Figs 2.13(d) and 2.13(f)), the greater is the range of movement achieved for a given angular displacement of the joint or range of muscle shortening. Thus, as will become evident in subsequent chapters, the mechanical organization of muscles in relation to an animal's skeleton greatly influences how animals adjust to scale effects of size, their running speed, their ability to accelerate, and their ability to maneuver.

The fiber architecture of muscles is also often related to their mechanical advantage and organization within a limb. Muscles which attach further from a joint's rotational axis typically have longer fibers. This enables them to maintain a fractional shortening range similar to that of muscles with shorter fibers

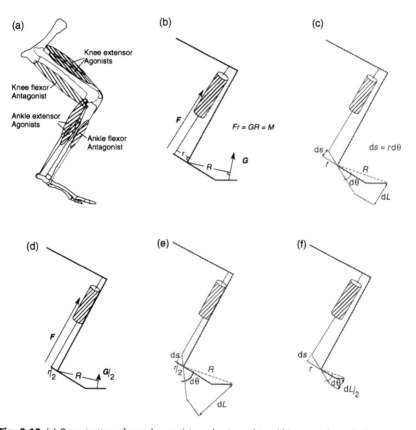

Fig. 2.13 (a) Organization of muscle agonists and antagonists within a vertebrate limb.
(b) Schematic diagram showing the moment (M) balance of an extensor muscle force (F) relative to
the ground reaction force (G) at the ankle joint, which depends on the mechanical advantage, or
moment arms (r and R), of each force vector. This moment balance ignores moments due to
segment inertia and weight, which are quite small in relation to the moment produced by the
ground reaction force (G) during limb support. (c) Diagram showing how the mechanical advantage,
or 'in-lever' of a muscle affects joint motion and limb displacement (dL). The angular motion ($d\theta$) of
a distal segment (in this case, the foot) depends on muscle shortening (ds) relative to its moment
arm (r). (d) Diagram showing how the moment that a muscle (ankle extensor) can produce is
reduced by half when its moment arm is reduced by 50 per cent (i.e. the muscle acts closer to the
joint). (e) By acting closer to the joint, or with a shorter 'in-lever' (moment arm $= r/2$), shortening of
the muscle (ds) produces twice the range of motion (and a twofold increase in joint velocity).
(f) With a shorter 'out-lever' (defined as the length of the distal segment, or foot, in this case), the
range of motion is reduced for a given joint angle displacement compared with (c). These differences
in muscle mechanical advantage reflect a fundamental trade-off between force (or moment) and
velocity of movement, as well as the range of motion that a muscle can produce. This affects
locomotor performance and the arrangement of muscles with respect to the skeleton in animals.

which act closer to the joint. This is advantageous because of the force–length property of striated muscle fibers discussed above (section 2.1.2). By having muscles which act close to and more distantly from a joint, an animal can also achieve muscle gearing, much like the transmission of an automobile. The former provide a high gear for fast, but less forceful, movements (Figs 2.13(c) and 2.13(e)), and the latter produce a low gear for slower more forceful movements (Figs 2.13(b) and 2.13(d)). Muscles which act in concert to produce a similar joint movement are termed 'agonists' or 'synergists' and are thus opposed by an antagonist muscle or muscle group.

Finally, the range of motion at a joint is largely dictated by the shape of the joint. Most exoskeletal joints are hinged with a single degree of freedom, i.e. rotation can only occur about a single axis (thus these joints have two sets of muscle antagonists—flexors and extensors). The joints of vertebrate endoskeletons, on the other hand, range from hinged joints with a single degree of freedom to ball-and-socket joints with three degrees of freedom in which rotation can occur about three independent axes. With a greater range of motion, the number of antagonist muscle pairs required to produce and control reciprocal joint motion is necessarily increased. Generally, multi-axial joint motion is characteristic of more proximal joints (the shoulder and hip joints of vertebrates), whereas distal limb joints are more restrictive (one and two degrees of freedom) in their potential range of motion. Possible advantages of this organization of limb joint design are that more distal segments of the limb can be made lighter (fewer sets of antagonistic muscles are required to control joint motion) and more easily controlled by the nervous system (Chapter 10).

2.3   Summary

The properties of muscles and skeletons relevant to the mechanics and energetics of locomotion have been reviewed in this chapter. We can now use the basic principles that have been discussed here, and which are broadly shared across a diversity of animal taxa, to understand the underlying function of the musculoskeletal system in various modes of locomotion. These will be discussed in the following chapters. The energetic aspects of movement which relate to how muscles generate force and do mechanical work, how the skeletal system transmits force, and how tendons store elastic energy will then be discussed. We shall conclude with a consideration of how the nervous system integrates sensory information to control the activation and recruitment of muscles involved in locomotive movements.

3 | Movement on land

Terrestrial locomotion requires that animals overcome gravity to support and move their weight and accommodate to changes in substrate and terrain. Animals generally accomplish this by propelling themselves with limbs which exert forces on the ground. In some instances animals have disposed of limbs (snakes) or never evolved them in the first place (worms) and rely instead on contractions of their axial musculature to transmit force between their body axis and the ground. Undulatory modes of terrestrial locomotion are nearly always associated with a burrowing existence. In other animals (salamanders and lizards), some combination of body undulation and limb propulsion is used to move the body forward. However, the majority of terrestrial animals rely primarily on limbs for movement. In this chapter we shall discuss the mechanisms and strategies for legged locomotion on land.

3.1 Biological wheels: why so few?

Interestingly, man's own device for efficient land transport—the wheel—has rarely evolved in biology as a means of terrestrial locomotion. Other than for a few organisms (most notably tumbleweed), wheels have not proven a viable solution for animal transport (LaBarbera 1983). In part this reflects the difficulty of evolving a biological rotary engine, at least beyond a molecular level of organization (see Chapter 4 on bacterial flagellar locomotion as an example in which rotary motion for movement has evolved). However, a basic reason is that wheels simply are not a good design for movement over uneven terrain. (Over the course of human history, wheel-based transport has required substantial infrastructure costs associated with building and maintaining level roads, and the frequently encountered pot-hole attests to the poor performance of a wheel when negotiating an uneven surface.)

Instead, natural selection has favored the evolution of limbs to negotiate variable terrain. The ability of animals to maneuver around obstacles and over uneven surfaces is unparalleled by any man-made vehicle. Because of this, recent robotics research has turned to the design of animal limbs and legged transport as a means for better understanding and designing robotic vehicles capable of negotiating natural terrain (Raibert 1986; Full and Koditschek 1999).

3.2   Limbs as propulsors: support and swing phases

When a terrestrial animal's limb contacts the ground it exerts, and thus experiences, a 'ground reaction force'. The vertical component G_V of the ground reaction force serves to support the animal's weight, while the horizontal fore–aft and mediolateral components G_H and G_{ML} allow the animal to accelerate or decelerate and to maneuver and balance (Fig. 3.1). Over a series of strides the average vertical force exerted on the ground by the limbs must equal an animal's weight. This can be defined mathematically as the impulse $\int G_V \, dt = W$. As shown in Fig. 3.2, this requires that the shaded areas during which $G_V < 0.5 \, W$ (assuming that each limb of a biped or pairs of limbs of a quadruped support half the animal's weight) are matched by the summed hatched areas during which $G_V > 0.5 \, W$. At rest, the force acting on a limb is approximately equal to W/n, where n is the number of limbs that support the

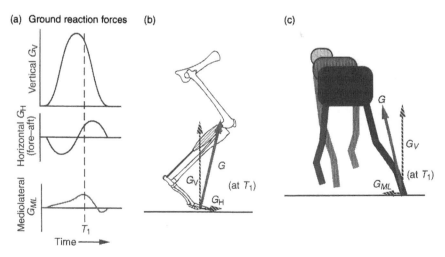

Fig. 3.1 (a) Representative vertical, horizontal fore–aft and mediolateral components of the ground reaction force exerted on a limb during the support phase of the stride. (b) Lateral view showing vertical and fore–aft components. (c) Posterior view showing vertical and mediolateral components acting on the limb at time T_1.

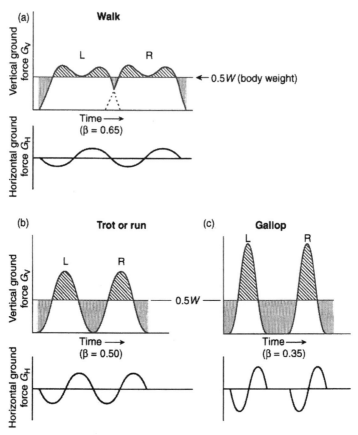

Fig. 3.2 Representative vertical and horizontal fore–aft ground reaction force components for (a) a walk, (b) a trot or run and (c) a gallop. The level of force required to support the body weight of a quadruped through time is 0.5 W (for a biped it would be W). For each gait, the ground force must rise above this level for a period of time to offset the time during which the ground force is less than 0.5 W for a quaduped (or W for a biped). This is depicted by the hatched and shaded regions of the vertical force records in each case. For slower gaits (with greater duty factor β), the time of limb support is longer, requiring a smaller maximum vertical ground force.

animal's weight ($n = 2$ for a biped and $n = 4$ for a quadruped). The exact distribution of weight support depends on the location of the animal's center of mass relative to its limbs. When an animal moves the forces exerted by the limbs on the ground rise and fall during limb support, and are zero whenever no limbs are on the ground (defined as the 'aerial phase' of a stride). As a result, the maximum forces exerted on the ground by a single limb are always much higher than those sustained when an animal is standing at rest. If the limbs are kept on the ground for a longer period of time, smaller forces are required

(Fig. 3.2(a)), but this limits the speed of movement. To move faster, animals must move their limbs more rapidly, reducing the time of limb contact with the ground and thereby increasing the magnitude of force that must be generated against the ground (compare Figs 3.2(b) and 3.2(c) with Fig. 3.2(a)). Whereas peak ground forces acting on an individual limb may be less than body weight when an animal moves slowly, they can be much greater than body weight when an animal moves at high speed. The maximum force required by a single limb can also be reduced by using more limbs to support an animal's weight (e.g. a hexaped versus a quadruped).

The locomotor cycle, or stride, can be divided into the support and swing phases of each limb, with the stride period being the time required to complete one cycle of limb movement (Fig. 3.3(a)). These are also often referred to as the propulsive and recovery phases of the limb. The relative fraction of the stride period T_s represented by a limb's support or ground contact phase t_c is defined as the limb's duty cycle ($\beta = t_c/T_s$). Hence, as animals

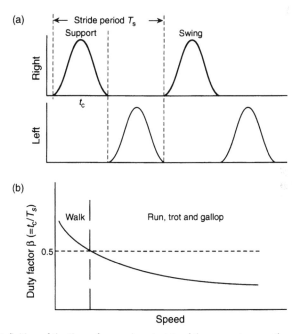

Fig. 3.3 (a) Definition of the time of ground contact t_c of the support versus the swing phase of a stride. The duty factor β is the ratio of limb contact time t_c to stride period T_s. (b) The duty factor declines with increased speed. This is because the fraction of the time that the animal supports its body weight on the ground during each stride decreases because of the need to move the limbs more quickly. When β falls below 0.5 animals typically switch to a running or trotting gait. At very fast running, trotting or galloping speeds, β may decrease to as low as 0.2 (or 20 per cent of the stride period T_s).

move faster the duty factor of their limbs decreases (Fig. 3.3(b)), requiring an increase in the maximum forces exerted against the ground. Animals move faster by increasing either their stride frequency or their stride length (the distance traveled over one stride cycle), or by doing both. The relative importance of the two depends on the type of gait employed by the animal. Even when an animal moves at steady speed, its limbs exert decelerating and accelerating horizontal forces in the direction of travel during each phase of limb support (Figs 3.2 and 3.4). The relative magnitude of acceleration versus deceleration differs according to which limb (i.e. fore, middle or hind) is in contact with the ground, and may vary according to the animal's size, speed and gait (Figs 3.4(a) and 3.4(b)). The net horizontal accelerating

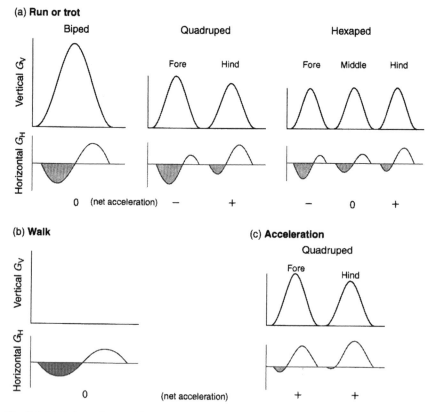

Fig. 3.4 (a) Horizontal ground forces for steady speed running of a biped (e.g. human), a quadruped (e.g. squirrel) and a hexaped (e.g. cockroach) compared with the ground forces exerted during (b) walking and (c) acceleration. The relative importance of deceleration (shaded portion of horizontal force curves) versus acceleration for the different limbs is shown in each case. See text for further details.

force exerted by an animal's limbs is positive when an animal accelerates to increase its speed (Fig. 3.4(c)), negative when it slows down and zero when it moves at a steady speed (Fig. 3.4(a)). Animals also exert mediolateral forces with their limbs (Fig. 3.1), particularly when changing direction (see section 3.4), but these are typically smaller in magnitude than those exerted in the vertical and horizontal directions when they are moving in a straight direction.

3.3 Limb mechanical advantage and joint moments: interaction of limb posture and ground reaction force

The ground reaction force acting on an animal's limb relative to the limb's posture during support is the main determinant of the moments, or torques, developed at the joints of the limb. In addition to these 'external joint moments', the inertia and weight of individual limb segments also contribute to the net joint moment. Although important during the swing phase of the limb, these are generally quite small relative to the 'external' moments developed due to the ground reaction force (Alexander 1983; Winter 1990). By tracking the movements of an animal's limb and the position of its joints with respect to the time course of the ground reaction force, it is possible to determine the external moments developed during limb support (Fig. 3.5). This is now commonly done by obtaining a high-speed video recording of limb kinematics synchronized to recordings of the ground reaction force components by means of a force platform. For humans and large animals, it is possible to attach reflective markers which special infrared video cameras can track for automatic collection of the coordinate data for analysis. Figure 3.5 shows an animal's limb position (in lateral view) and the ground reaction force acting on the limb, derived from the vertical and horizontal components (G_V and G_H), for three video frames. By synchronizing the ground force components with the coordinates of the joints, the joint moments developed in the plane of the animal's direction of travel can be determined. For more complex three-dimensional motion and joint moments, it is necessary to obtain one or more additional views of the limb and to incorporate the mediolateral component G_{ML} of the ground reaction force.

Because the moment acting at a joint must be balanced by muscle force F_m, the magnitude of this force is determined by the muscle's (or muscle group's) moment arm r relative to the moment arm R of the ground reaction force G (Fig. 3.6(a)), or $F_m \times r = G \times R$ (moments are the cross-product of the vectors of force and moment arm). Defining r and R as the perpendicular distance to the joint center and rearranging, gives the following relationship between

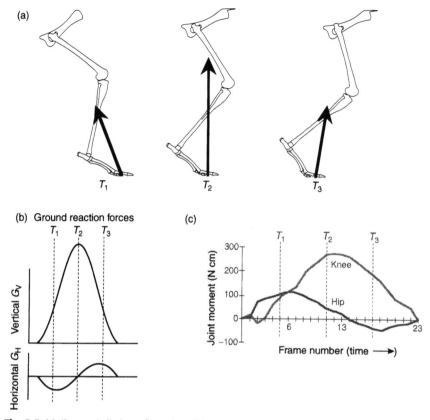

Fig. 3.5 (a) Changes in limb configuration with respect to the net ground reaction force G for three frames (times T_1, T_2 and T_3) corresponding to the vertical and horizontal ground forces shown in (b). (c) Changes in ground reaction force moments acting at the hip and knee versus time. G exerts a flexor moment at the knee throughout most of limb support (requiring knee muscle extensor force to counter this). G also exerts a flexor moment at the hip (requiring hip extensor force) during the first 60 per cent of limb support but, as G passes behind the hip joint, it exerts an extensor moment during the latter 40 per cent of limb support (which must be balanced by hip flexor activity).

mechanical advantage and force: $r/R = G/F_m$. These moment arm and force ratios provide a measure of the **effective mechanical advantage** (EMA) of limb muscles. When determined over the entire time period of support this can be calculated as

$$\text{EMA} = \int G \, dt / \int F_m \, dt. \tag{3.1}$$

Defined in this way, the EMA of a limb represents a measure of the magnitude of time-integrated muscle force that must be developed for a

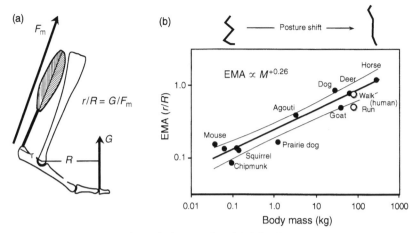

Fig. 3.6 (a) Effective limb mechanical advantage (EMA) is defined as the ratio r/R of the moment arms of the ground reaction force G relative to the extensor muscle force F_m. (b) Because of changes in locomotor limb posture, EMA increases with increasing size for terrestrial mammals, scaling as $M^{0.26}$ (Biewener 1989). This allows larger mammals to support their weight while running without exceeding the force capacity of their muscles and the strength of their limb bones.

given time-integrated ground force during the support phase of the stride, i.e. the ratio of ground impulse to muscle impulse. By aligning the joints more closely in the direction of the ground reaction force, the magnitude of joint moments and hence muscle forces can be reduced (Biewener 1989). This has important implications for both the mechanical design of animal limbs and the energetic cost of locomotion. The latter will be discussed in Chapter 9.

Changes in limb mechanical advantage are important for enabling different-sized mammals to maintain similar peak muscle and bone stresses, and thus similar safety factors to failure. As we noted in Chapter 2, area-to-volume scaling poses a significant design problem for achieving comparable mechanical support over a broad range of size in animals built of similar tissues. In mammals and birds (and other vertebrates), the capacity of skeletal muscle to generate force per unit area (muscle stress) and for limb bones to transmit mechanical load per unit area (bone stress) are very similar. Consequently, if limits to locomotive performance are to be avoided, peak muscle and bone forces must scale in proportion to changes in muscle fiber area and bone area, and not in proportion to body mass.

Different-sized mammals achieve a mass-specific reduction in peak musculoskeletal forces by adopting different locomotor postures, which changes the mechanical advantage of their limbs (Biewener 1989) (Fig. 3.6(b)). Compared with the crouched postures and poor mechanical advantage of small mammals, larger mammals run with more upright postures which align their joints more closely with the ground reaction force, enabling them to operate with a greater mechanical advantage. This reduces the magnitude of force that their muscles and bones must support relative to their weight. Because of this, muscle and bone forces scale as $M^{0.74}$ ($\propto M/EMA$) rather than as $M^{1.0}$, matching closely the scaling of muscle fiber area and bone area (which scale as $M^{0.75}$). Consequently, muscle and bone stress are maintained nearly constant across a wide size range within terrestrial mammals.

While important to maintaining similar skeletal safety factors, a size-related shift in posture probably represents a trade-off in maneuverability at larger size. Changes in mechanical advantage within terrestrial mammals achieved through shifts in limb posture are also probably limited for species greater than 300–500 kg in size (Biewener 1990). Consequently, very large terrestrial animals have even more limited locomotive performance with respect to speed. Interestingly, within a given species limb mechanical advantage does not seem to change much as a function of speed and gait. This largely reflects the consistent pattern of limb kinematics, muscle recruitment and particular musculoskeletal organization of an animal's limb.

Whether or not size-dependent changes in locomotor posture occur in other terrestrial vertebrates or in invertebrate runners, it seems likely that limb posture plays an important role in dictating the forces that muscles must develop to support gravitationally induced ground reaction loads. Certainly, changes in posture are important to how animals maneuver and achieve balance and stability, in addition to how it affects musculoskeletal loading.

3.4 Locomotor gaits

Locomotor gaits are generally defined by the relative timing of support among the limbs of the animal during the stride. Changes in gait are associated with movement at different speeds and typically involve a discontinuous change in limb kinematics and/or the mechanics of support. Three general classes of gaits have been defined: walking, running or trotting (and hopping) and galloping. More detailed descriptions and comparisons for vertebrate animals are presented elsewhere (Gambaryan 1974; Hildebrand 1988).

3.4.1 Walking

Walking gaits (Fig. 3.7) generally involve overlapping periods of support among the limbs (duty cycles $\beta > 0.5$). For quadrupedal animals (four-legged vertebrates), hexapedal animals (six-legged insects) and octapedal animals (eight-legged crustaceans or spiders), this means that walking incorporates periods during which three or more limbs are in contact with the ground,

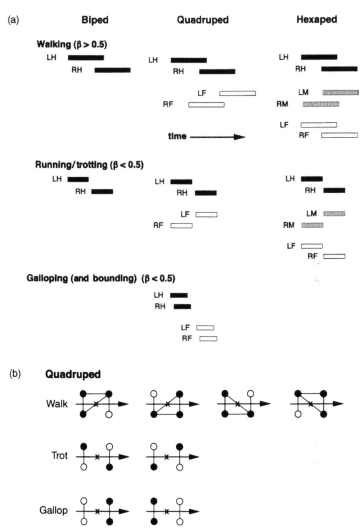

Fig. 3.7 (a) Gait diagrams for a biped, a quadruped and a hexaped using different gaits:LF, left forelimb; RF, right forelimb; LH, left hindlimb; RH, right hindlimb; LM, left middle limb; RM, right middle limb). The timing and duration of foot support periods is shown by the rectangles (β is the duty factor). (b) Representation of the pattern of limb support for a quadruped moving to the right (arrow) for each gait. Solid circles depict the limb's support phase and open circles the limb's swing phase (the cross represents the location of the animal's CM).

providing a statically stable base of support (Fig. 3.7(b)). Static equilibrium is achieved because the body's center of gravity falls within the triangular area of support represented by the limbs (analogous to a table). Although static equilibrium enhances stability, it can only be achieved at very low speeds.

Consequently, over most of their speed range, animals must rely on dynamic balancing mechanisms. Bipedal walking (used by humans and ground-dwelling birds) also involves overlapping periods of limb support, but because only two limbs are involved static equilibrium is rarely achieved. As with faster gaits, dynamic equilibrium must be relied on for balance and stability.

Animals can change speed within a gait, but to move over a greater range of speed they must change gait. When an animal changes gait from a walk to a trot or a run, not only is its stride period reduced owing to the increase in stride frequency required to move at a faster speed, but the relative fraction of limb support (duty factor) also decreases (and swing phase increases). By decreasing the fraction of the stride period that a limb contacts the ground, the overlap among the alternating support phases of the limbs is reduced.

3.4.2 Trotting and running

Trotting and running gaits are typically characterized by limb duty cycles below 0.5. Consequently, there are no overlapping support periods between alternating support limbs. In a quadrupedal trot, for instance, the diagonal forelimb and hindlimb move in unison (i.e. are 'in phase'), contacting the ground at the same time and leaving the ground to begin their swing phase before the contralateral forelimb and hindlimb contact the ground to begin their respective support phases (Fig. 3.7). Bipedal human running similarly involves the absence of an overlap between the support phases of the two limbs. However, in avian bipeds the gait transition from walking to running is less distinct. Avian bipeds run with an overlapping support phase between their two limbs over much of their speed range. Only at the fastest running speeds do birds have non-overlapping support. Their relatively longer feet may account for some of this difference, but other factors also likely play a role.

Hexapedal insect running (Full and Tu 1991; Full *et al.* 1991) involves a similar gait pattern to that observed in vertebrates, in which three limbs (fore- and hindlimb on one side and the contralateral middle limb) are used in alternating 'in-phase' tripods of limb support, equivalent to the diagonal fore- and hindlimb of a quadrupedal trotting mammal or the alternating hindlimbs of a bipedal runner (Fig. 3.7). Sideways running crustraceans (Blickhan and Full 1987) also appear to adopt a similar strategy. In the case of ghost crabs, pairs of the four leading (limbs 2 and 4 and limbs 3 and 5) are used in combination with alternating pairs of its trailing limbs to achieve overall patterns of limb support which produce forces and motions of the body (see section 3.8.2) characteristic of a bipedal running gait.

Two other gaits similar to running and trotting deserve mention; these are hopping and pacing (not shown). Hopping is used by several marsupial species (kangaroos, wallabies, rat kangaroos) and rodent species (kangaroo rats, jerboas, spring hares), as well as by many frogs and toads. (Hopping in anurans often involves forelimb contact and is not as continuous or steady as the hopping gait of mammals, which is often described as being 'richochetal'. Hence it will be considered in Chapter 7 when jumping is discussed.) During hopping the two hindlimbs move in phase rather than in an alternating fashion as for bipedal runners. Pacing involves the use of in-phase limb support by the ipsilateral fore- and hindlimbs of quadrupeds, rather than the contralateral diagonal fore- and hindlimbs of trotting animals. Although horses can be trained to pace for racing competition, it is an unnatural gait. Few quadrupeds pace under normal circumstances (camels are a notable exception), because it is a less stable gait owing to its ipsilateral pattern of limb support, which results in a greater rocking motion of the body compared with trotting.

All these gaits are similar in that they involve a bouncing spring-like motion of the body on the supporting limbs (see Fig. 3.11 below and section 3.8.2). Whereas bipedal runners typically increase running speed by increasing both stride frequency and stride length, quadrupedal trotters generally favor an increase in speed by increasing stride frequency and bipedal hoppers by increasing stride length (see Fig. 3.10 below). In all cases, the duty cycle of the limb decreases at faster speeds, requiring an increase in musculoskeletal forces.

3.4.3 Galloping

In addition to walking and trotting, quadrupedal mammals have evolved an additional gait, the gallop (sometimes referred to as a canter when used at lower speeds), to increase speed beyond that which can be achieved at a trot. The transition from a trot to a gallop involves a relative shift in the support phases of the fore- and hindlimbs, such that the two forelimbs move more or less in phase, followed by the two hindlimbs (Fig. 3.7). By shifting the phase of limb support to allow the fore- and hindlimbs to act together as pairs, galloping animals are able to increase their stride length to a greater extent than is possible by the rotational movement of the limbs alone. This is achieved by flexion and extension of the spinal column and rotation of the shoulder girdle and pelvis, which can increase stride length considerably. This is most dramatic in the pursuit gallop of a cheetah, and cats in general, in which spinal flexion and extension may provide as much as a 20 per cent increase in the animal's stride length. Spinal flexion and extension also occurs in smaller rodents and other carnivores, but is modest or absent in larger ungulates (horses, antelope, wildebeest, etc.) owing to scale constraints of size. A rigid backbone in large ungulates is necessary for effective support of the trunk, which is suspended between the hip and shoulder joints, and the need to transfer load between the fore- and hindlimbs during a gallop. This results from the asymmetric timing of the fore- and hindlimb support phases, in addition to the asymmetric pattern of deceleration (by the forelimbs) and acceleration (by the hindlimbs) that each set of limbs exert on the ground. Typically, increases in speed at a gallop mainly involve increases in stride length with little increase in stride frequency (see Fig. 3.10 below).

At a slow gallop (or canter), one forelimb lands slightly ahead of the other forelimb, followed by a similar pattern of support by the two hindlimbs. Most often, the phase difference between the forelimbs is greater than that between the hindlimbs. At faster galloping speeds, the two forelimbs and two hindlimbs progressively land more in phase with one another and limb duty cycles decrease (to as low as 0.2). When the two hindlimbs move fully in phase, the gait is defined as a 'half-bound'. When the forelimbs and hindlimbs each move together in phase, the gait is considered a 'full-bound'. Bounding gaits are typical of smaller rodents and carnivores, as well as hares and rabbits. Indeed, these smaller animals seldom trot, commonly changing gait directly from a walk to a half-bound or bound even at low to moderate speeds. Because of the reduced limb duty cycle, galloping involves aerial phases which may intervene between one or both sets of limb support phase. These aerial phases are a necessary consequence of the increased stride length that the animals achieve to increase their speed at a gallop.

3.5 Maneuverability versus stability

As noted above, by providing a tripod of support, walking gaits can provide static stability. Although these gaits enhance stability, they do not allow for rapid movement and often limit an animal's maneuverability. Maneuverability at faster speed requires dynamic stability. Even tortoises, upon which the virtues of slow and steady walking have been the fancy of children's tales, depend on dynamic stability during much of the time that they walk. Much like learning to ride a bicycle, forward momentum and dynamic exchanges in kinetic and potential energy of the body (see section 3.8.2) provide mechanisms for stabilization of the body at faster speeds. Such stabilizing mechanisms also depend on the control of movement achieved by the nervous system. Nevertheless, a considerable degree of dynamic stability appears to be engineered into the basic mechanical design of animals and their limbs. Disturbances in balance or support when animals move over uneven and less predictable terrain can be accommodated by energy absorption at particular joints of a supporting limb, and compensating energy production at other limb joints, to provide dynamic stabilization of the body. While the underlying principles and mechanisms of dynamic stabilization are only now being recognized and appreciated, they are key to understanding how mechanical design can simplify the seemingly complex task of control by the nervous system (see Chapter 10).

Size-related trade-offs between maneuverability and stability also occur. To make a turn, an animal must generate a laterally directed force with its limb that is resisted by a medially directed component of the reaction force exerted on the ground (G_{ML}) (Fig. 3.8(a)). However, in doing so, the animal must avoid toppling over (laterally) and falling. This is determined by the angle θ of the animal's limb with respect to the ground and the magnitude of the toppling moment M_{top} ($=G_{ML}L\sin\theta$) generated during a turn relative to the animal's weight (Fig. 3.8(a)). To avoid toppling over the animal must satisfy

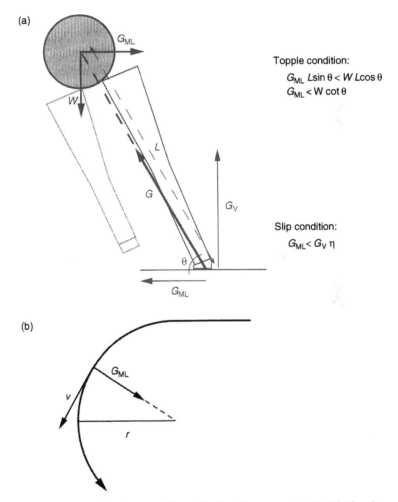

(a)

G_{ML}

W

L

G

G_V

θ

G_{ML}

Topple condition:

$G_{ML}\,L\sin\theta < W\,L\cos\theta$
$G_{ML} < W\cot\theta$

Slip condition:

$G_{ML} < G_V\,\eta$

(b)

G_{ML}

v

r

Fig. 3.8 (a) Mechanics of turning to avoid toppling (toppling moment $= G_{ML}L\sin\theta$) and to avoid slipping. (b) Toppling and slipping in relation to turning radius and speed. See text for further details.

the following condition:

$$G_{ML}L\sin\theta < WL\cos\theta \tag{3.2}$$

where L is the distance to the animal's center of mass from the base of its supporting limb. Equation (3.2) can be simplified to

$$G_{ML} < W\cot\theta \tag{3.3}$$

or

$$W < G_{ML}\tan\theta. \tag{3.4}$$

Therefore, by bringing their centers of mass closer to the ground and lowering θ, animals can make sharper turns. This is reflected by the more crouched or flexed limb posture that animals adopt when making a turn. In general, smaller animals can make sharper turns than large animals because they run with more crouched postures. In addition, because their shorter limbs require that they take more steps to move a given distance, smaller animals can turn over shorter distances, allowing them to turn more rapidly as well.

In addition to toppling, animals must also avoid slipping when making a turn. This requires that a second condition be met:

$$G_{ML} < G_V\eta \tag{3.5}$$

where η is the coefficient of friction between the animal's foot and the ground and G_V is the vertical component of the ground reaction force. The radius r of a turn will also depend on how fast the animal is traveling and its ability to avoid slipping (Fig. 3.8b)). For a given turning radius, the medial component of the ground force is

$$G_{ML} = mv^2/r \tag{3.6}$$

which, in order to avoid slipping, requires that

$$\eta > k(v^2/gr) \tag{3.7}$$

where v is the animal's velocity tangent to the turn, r is the radius of the turn, g is the gravitational acceleration constant and k is the ratio of the magnitude of the vertical ground reaction force to an animal's weight ($G_V = kW = kmg$; recall from section 3.2 that G_V may be greater or less than an animal's weight, depending on its speed and gait). In addition to the surface properties of the foot and the ground, η will depend on the relative area of contact of the foot with the ground. Consequently, because smaller animals generally have proportionately larger feet (i.e. greater surface area of contact) for their weight, η will tend to vary inversely with body size. Therefore smaller animals are less likely to slip than larger animals. Reductions in foot (or hoof) size, as cursorial adaptations for running fast, come at the expense of reducing η. The greater maneuverability and stability of smaller animals is apparent to anyone who has watched a larger animal chase its smaller intended prey (for example, when Fido chases a squirrel to the nearest available tree). Almost always, in such instances, maneuverability more than compensates for the slower speed of the smaller animal, enabling it to escape its pursuer.

3.6 Stride frequency and stride length versus speed and size

Size influences many general features of animal locomotion, in addition to maneuverability and stability. Classical arguments of isometric scaling made by A.V. Hill (1950), the Nobel laureate who received recognition for his work on muscle energetics, suggest that small and large animals should achieve similar top speeds. This is based on the notion that length scales in inverse proportion to frequency ($l \propto f^{-1} \propto W^{0.33}$) (see Chapter 1). Small animals with short limbs swing their limbs at high frequencies, whereas large animals with long limbs move them at lower frequencies. Consequently, these arguments suggest that the maximum velocity $v = L_s F_s$, where L_s and F_s are stride length and stride frequency, of similarly constructed animals should be constant and independent of size.

However, when compared across a large size range, terrestrial birds and mammals exhibit a scaling of stride frequency proportional to $W^{-0.14}$ and a scaling of stride length proportional to $W^{0.38}$ (Heglund and Taylor 1988) (Fig. 3.9). These scaling relationships were obtained by comparing stride frequencies and stride lengths of different-sized animals at equivalent points of gait (e.g. each species' preferred trotting speed or trot–gallop transition speed). Because stride length increases more rapidly than the decrease in stride frequency with size, larger animals are generally able to attain greater top speeds. Certainly two of the fastest land animals, the cheetah (maximum sprint speed, ~100 km/h) and the pronghorn antelope (maximum sustainable speed, ~65 km/h) are at the large end of the size spectrum of terrestrial animals. Comparable data are unavailable for invertebrate runners, but it seems likely that similar patterns may be found. However, at very large sizes (>500 kg) locomotor performance of terrestrial mammals is constrained due to the effects of mechanical stress on musculoskeletal design (see section 3.3), and variables such as speed (Garland, 1983) and maneuverability decline. It is unlikely that similar constraints of size affect performance in the largest terrestrial invertebrates because their size is limited more by the mechanical (local buckling) and growth constraints of their exoskeletons.

Increases in locomotor speed within and between gaits show differing patterns relative to changes in stride frequency and stride length that are taxonomically and size related. Avian and human bipeds increase their speed

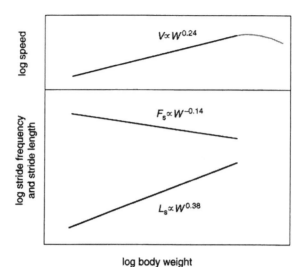

log body weight

Fig. 3.9 Scaling of maximum running speed v, stride frequency F_s and stride length L_s in birds and mammals ($v = F_sL_s$).

by increases in stride frequency and stride length at both a walk and a run, whereas hopping bipeds increase their speed almost entirely by increasing stride length, with stride frequency held nearly constant (Fig. 3.10). Small and large mammalian quadrupeds also differ in the pattern of increases in speed due to stride length and frequency versus gait. Small mammals (e.g. squirrels) increase speed mainly by increasing speed at a gallop. Increases at a walk and a trot are more restricted, involving both increases in stride length and stride frequency. At a gallop, increased speed is mainly achieved by an increase in stride length similar to the pattern observed in hopping bipeds. In larger quadrupeds, a greater fraction of the animal's speed range is achieved within a trot, with increases in both stride frequency and stride length contributing to speed increase. Once the animal changes gait to a gallop, increases in stride length contribute most of the increase in speed.

Non-mammalian quadrupeds (i.e. salamanders, lizards and alligators) generally increase speed by comparable increases in stride frequency and stride length. This pattern of speed increase is generally characteristic of both walking and trotting gaits. However, as these animals approach their fastest speeds, increases in speed appear to be achieved progressively more by increases in stride length than by stride frequency. Patterns of speed increase associated with changes in stride length and stride frequency as a function of gait have been much less well studied in insects and other invertebrate runners. In the few invertebrate species that have been studied (Fig. 3.10), increases in speed are achieved by increases in stride frequency and stride length at slower speeds

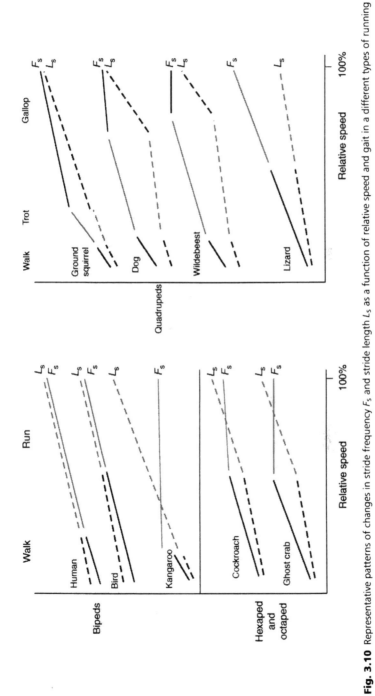

Fig. 3.10 Representative patterns of changes in stride frequency F_s and stride length L_s as a function of relative speed and gait in a different types of running and hopping (kangaroo) animals.

but depend more on an increase in stride length at faster running speeds (Ting *et al.* 1994), similar to the pattern observed for mammalian quadrupeds.

3.7 Mass–spring properties of running

The fairly uniform stride frequency that is maintained during bipedal hopping and quadruped galloping has led workers to speculate that this represents a resonant frequency of the animal's body, which depends on the animal's mass and the overall spring stiffness of its limbs. The resonant frequency of any simple mass–spring system can be described by the following relationship

$$f_{nat} = (k/m)^{0.5}. \tag{3.8}$$

Thus, for a given spring stiffness k, larger masses vibrate at lower resonant frequencies, consistent with the scaling decrease in stride frequency noted above for terrestrial mammals and birds. For $f_{nat} \propto W^{-0.14}$, this relationship predicts that the overall stiffness of the animal's limbs and body scales as $W^{0.72}$. Thus larger vertebrate animals move with stiffer limbs and bodies than small animals, consistent with their reduced flexion of the vertebral column at a gallop. By galloping or hopping at their body's natural frequency, animals are believed to reduce their energy expenditure (Heglund and Taylor 1988).

The notion that terrestrial locomotion for many animals using a running, hopping or galloping gait involves spring-like function of the limb in support of the animal's body weight has led workers to model the limb and body as a simple mass–spring system (Fig. 3.11(a)) (McMahon *et al.* 1987; McMahon and Cheng 1990). The properties of a leg spring supporting the weight or mass of the body predict well the mechanical action and kinematics of limb and body movement of many terrestrial mammals and birds. A simple mass–spring model, such as that shown in Fig. 3.11(a), also predicts the scaling of increased leg-spring stiffness with size (Farley *et al.* 1993). Part of the change in leg stiffness can be explained by the more upright locomotor posture that larger mammals adopt in order to increase their limb mechanical advantage for weight support (section 3.3 and Fig. 3.6). It is also becoming apparent that the neural control of muscle activation during running is linked to controlling muscle stiffness and its effect on the spring-like behavior of the limb as a whole (see Chapter 10).

The observation that mechanical factors are linked to, and thus influence, animal gaits has led to the suggestion that such factors may provide a 'trigger', signaling an animal to change gait at an appropriate speed. Consistent with this hypothesis, when horses carry a load equal to 15 per cent of their weight, they change gait from a trot to a gallop at a lower speed than when they move unloaded (Farley and Taylor 1991). We shall see in Chapter 8 that changes

(a)

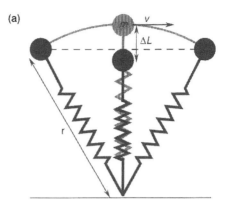

(b)

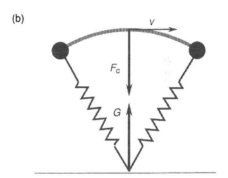

Fig. 3.11 (a) Mass–spring and (b) rotational behavior of body movement over an animal's supporting limb. This simple model has been applied successfully to explain many features of the spring-like behavior of terrestrial running gaits. It consists of a linear 'leg spring' that supports the mass of the running animal's body. Displacements ΔL of the whole-leg spring are determined by the spring stiffness of the leg relative to the magnitude of ground force. Motion of the body (v) reflects the centripetal motion of the body's mass over the leg spring. See text for further details.

in gait also influence the energy cost of locomotion and are important to the economy of an animal's locomotor activities.

3.8 Froude number and dynamic similarity

Observations of the mechanics and kinematics of limb support have also allowed gaits to be defined by a dimensionless factor known as the Froude number (Alexander 1989):

$$\text{Fr} = (v/gl)^{0.5} \tag{3.9}$$

where v is the animal's velocity, g is the gravitational acceleration constant and l is a characteristic length (such as hip height) of the animal. The Froude number normalizes the forward velocity of a moving animal to its limb length and gravitational acceleration. These parameters represent fundamental force interactions of stepping locomotion, in which the centrifugal force ($F = mv^2/r$) acting on the body's mass, as it rotates over a supporting limb, balances the ground reaction force acting on the limb from below (Fig. 3.11(b)). At the same

Froude number, geometrically similar animals move in a dynamically similar fashion. For example, two pendulums of different lengths, swinging through the same angle, move in a dynamically similar fashion. The Froude number also represents the ratio of a moving body's kinetic energy ($mv^2/2$) relative to its potential energy (mgl). Therefore equal Froude numbers imply equal ratios of kinetic to potential energy when an animal moves. Animal gaits (principally mammals) are fairly well defined by Froude number, with animals generally changing gait from a walk to a trot at Fr values of 0.3 to 0.5, and from a trot to a gallop at Fr values of 2 to 3 (Alexander and Jayes 1983). Consequently, animals change gait from a walk to a run when the ratio of kinetic to potential energy is about 0.2 and from a trot to a gallop when the ratio is about 1.25. As we shall see below, this reflects the fact that kinetic energy fluctuations of an animal's body become increasingly important at faster speeds. Although the Froude number and the concept of dynamic similarity appear to work well in terrestrial gravity, by ignoring inertial forces which are also important in locomotion, they do not appear to hold as well when gravity is altered (Kram 1997).

3.9 Inferring gait and speed of fossil animals

Fossil trackways provide a historical record of the stride length and footfall pattern of the animals that made them. If the foot impression of the trackway is distinct enough to determine the species that made the track and reliable reconstruction of its limb skeleton is available from fossil material, it is possible, using an estimate of Fr based on stride length and the animal's limb height, to estimate the speed and gait of the animal at the time that it made the track. Analyses such as this have been carried out for fossil trackways of human ancestors and fossil dinosaurs (Alexander 1976, 1984). In the case of fossil hominid trackways made at Laetoli, Tanzania, 3 million years ago, hominids were estimated to be walking at speeds of 0.5–0.75 m/s. This is lower than the preferred walking speed of modern humans which ranges from about 1.0 to 1.3 m/s. The difference is as likely to reflect the 33 per cent shorter stature of our early ancestors rather than limitations of their bipedal gait.

In the case of dinosaurs, estimates from trackways located in Texas suggest that larger quadrupedal sauropod dinosaurs walked rather slowly at 1 m/s and that moderate-sized bipedal theropod dinosaurs walked at speeds close to 2.2 m/s, but ran at speeds as high as 12 m/s. This would certainly be a competitive speed for an Olympic 100-m race, over which human sprinters average 10 m/s! However, it is likely that the very large carnivorous theropod dinosaurs, such as *Tyrannosaurus*, were too large to run. Recent analysis (Hutchinson and Garcia 2002), based on estimates of the extensor muscle mass needed to support their body weight compared with the amount of muscle that modern running birds and cursorial mammals possess, indicate that very large theropods (about 6000 kg) were probably poor runners and did not exceed speeds of 8–10 m/s. Together with information about the width of an animal's stance and length of its step, the use of mechanical parameters such as Froude number and biomechanical estimates of muscle capacity, which can be obtained from extant animals and applied to fossil ones, enables a more in-depth analysis of skeletal form and locomotor function in fossil animals than would otherwise be possible based on skeletal morphology alone.

3.10   Mechanical work: potential and kinetic energy changes during locomotion

In addition to generating forces against the ground to support their weight, terrestrial animals must also exert forces to maintain the oscillations in potential energy (PE) and kinetic energy (KE) associated with moving their bodies. These energy fluctuations (of the animal's center of mass) are inherent in all forms of legged transport and exhibit surprisingly conservative patterns for differing locomotor gaits and modes (Cavagna *et al.* 1977) when compared across a broad diversity of terrestrial animals, including both arthropods and vertebrates. Because of this, patterns of energy change can also be used to define an animal's gait.

In all forms of legged locomotion, PE must be supplied by an animal's muscles to raise and lower its body's center of mass during every step. Similarly, KE must be supplied to re-accelerate both its center of mass and its limbs relative to its center of mass. This is because an animal's center of mass rises as it moves over a supporting limb and falls as the body's weight is transferred over to the opposite side limb (Figs 3.11 and 3.12). In addition, even when an animal is moving at a constant average speed, it experiences small fluctuations in forward speed (as well as vertical speed), slowing down when it first lands on the ground with one of its limbs and then speeding up to regain its forward speed during the latter half of limb support. This is reflected in the decelerating and accelerating horizontal components of the ground reaction force (Figs 3.1 and 3.4). The decrease in forward speed represents a loss in the KE of the animal's body that its muscles must regenerate, or must be recovered from elastic sources, in order to maintain a constant average forward speed. Changes in vertical KE also occur during each step owing to the vertical component of ground reaction force which decelerates the animal's 'fall' when it first lands and re-accelerates its center of mass upward to regain the PE lost during the fall. Finally, the oscillatory movement of the limbs relative to the body also requires KE to decelerate the limb at the end of its swinging motion (both at the end of swing and the end of support) and subsequently to move it in the opposite direction.

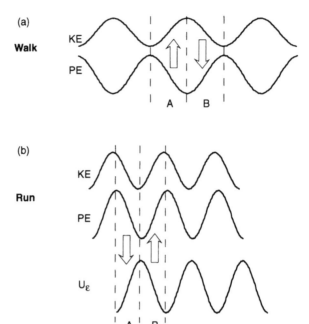

Fig. 3.12 Changes in the potential energy (PE) and kinetic energy (KE) of an animal's CM as a function of time for (a) walking and (b) running gaits. During walking, PE and KE fluctuate out of phase, allowing an exchange from KE to PE during time period A and an exchange from PE to KE during time period B. During running, PE and KE fluctuate in phase, preventing the exchange between each form of energy as during walking. Consequently, during running KE and PE are converted to elastic energy U_ε in spring elements of the limb during time period A and converted back to PE and KE during time period B. These energy exchange mechanisms reduce the amount of work that the muscles must perform, lowering the energy cost of terrestrial locomotion.

Losses in PE and KE require muscle work to sustain the forward speed and energy of an animal's body, and hence incur a metabolic cost for the work that the muscles must perform. In general, animals have evolved generally efficient and smooth modes of legged transport which minimize the oscillations in PE and KE of the body's center of mass (Fig. 3.12) and thus reduce the metabolic cost of the muscular work needed to maintain the body's energy equilibrium. In addition, animals have independently evolved similar mechanisms for reducing the work of locomotion by means of efficient exchange of PE and KE, or by elastic energy storage and recovery. Indeed, as we shall see in Chapter 9, metabolic patterns of energy use indicate that the energy cost of terrestrial locomotion is determined more by the magnitude and rate of muscle force generation than by the work that muscles perform. Hence mechanisms to reduce the energy fluctuations of the body and reduce the mechanical work of the muscles are important for economical gait.

3.10.1 Walking: body and limb movement as 'inverted pendulums'

All walking gaits involve an exchange of PE and KE of the body's center of mass (Fig. 3.12(a)). This is possible because fluctuations in PE and KE occur out of phase at a walk. PE is maximal at mid-support when an animal moves over its supporting limb (or limbs) and falls as the animal shifts weight support to the next supporting limb or set of limbs. As the animal 'falls' forward during this shift in limb support, its KE increases. Consequently, losses in PE at this time (time period A in Fig. 3.12(a)) can be exchanged into KE, reducing the amount of muscle work required to increase the KE of the animal's body. Similarly, as the animal's weight shifts over the next supporting limb(s), its KE decreases at the same time that its PE rises, enabling an opposite exchange of KE to PE (time period B). The ongoing exchange between PE and KE of the body's center of mass during walking gaits of bipeds, quadrupeds and hexapeds all occurs in a similar fashion, with an exchange of up to 70 per cent being achieved during walking (Heglund *et al.* 1982). Often this represents the animal's preferred, or optimal, walking speed and coincides with its ability to minimize its metabolic energy cost of transport (see Chapter 8, section 8.11). The exchange of PE and KE during walking is analogous to an 'inverted pendulum', based on the similar functional exchange of energy of the pendulum of a spring-wound clock which requires only a small amount of spring energy to be supplied during each tick to keep the clock running. In the case of walking animals, this energy is supplied by the muscles during each step and is decreased by the effectiveness with which the PE and KE of the body's center of mass can be exchanged. Similar exchanges in energy may also occur within and between the limbs, but these are much smaller components than those of the body as a whole.

3.10.2 Running, trotting, hopping and galloping: bouncing gaits

In contrast with walking, the PE and KE of the body's center of mass fluctuate in phase during running, trotting, hopping and galloping gaits (Fig. 3.12(b)). Therefore all of these gaits preclude significant energy conservation by means of exchange between PE and KE of the body. Does this mean that all of the energy lost during a step must be re-supplied by an animal's muscles, or is there some other means for conserving muscle work and metabolic energy? Rather than losing most of the PE and KE that is lost when landing on the ground, vertebrate runners absorb this energy at higher speed gaits (time period A in Fig. 3.12(b)) by storing it in elastic elements of the body, primarily the tendons or ligaments of the limbs. The elastic strain energy stored in these elements (see Fig. 3.14 below) is subsequently restored to the animal (time period B), increasing its PE and KE as it rebounds off the ground during the latter half of limb support. These elastic components underlie the spring-like properties that the limbs of animals exhibit when they run (Fig. 3.11). Similar to walking, the exchange between the PE and KE of the body's center of mass and elastic strain energy substantially reduces the amount of work that the muscles must do to keep the animal moving at a constant speed. Whereas the limb tendons are believed to be the main sites of energy savings for vertebrate bipedal running and quadrupedal trotting gaits, significant savings are also believed to occur within the ligaments and aponeuroses (muscle connective tissue sheaths) located in the trunks of quadrupeds when they gallop. However, the relative importance of energy savings by means of trunk versus limb elastic structures is less well known. Because larger quadrupeds gallop with stiffer trunks than smaller quadrupeds, it seems likely that elastic energy savings within the trunk relative to the limbs may be less important in larger animals.

Similarly, although there is evidence from whole-body mechanics that running in a hexaped (insect) or an octaped (crab) involves a bouncing gait, similar to that of terrestrial vertebrates, the sites and amount of elastic energy storage which occur in invertebrate runners are not well known at present. One problem is that the limb muscles of invertebrates transmit force via apodemes or directly to their exoskeleton, both of which are made of chitin. Because chitin is stiffer than vertebrate tendon, the amount of strain energy that can be developed for a given force is less (see Chapter 2). Nevertheless, as we shall see in Chapter 7, some invertebrates have evolved specialized catapult mechanisms for storing strain energy in their apodeme and cuticle that allow them to achieve impressive jump distances for their size.

3.11 Muscle work versus force economy

Because work must be done to swing the limbs back and forth and to raise the height of an animal's center of mass during every step, it was long thought that the main function of muscles in terrestrial locomotion was to perform the work necessary for these movements. Positive mechanical work is done by a muscle when it shortens as it develops force and 'negative work' is done when it is stretched (Chapter 2). These functions are certainly important when energy is required as an animal pushes off from the ground during running or jumping, and when energy must be absorbed when landing. However, as will be discussed in Chapter 9, the amount of work that different-sized animals perform to run at different speeds does not correlate well with the energy cost of terrestrial locomotion (Heglund et al. 1982). Instead, it appears that energy cost is determined more by the rate and magnitude of force that muscles must generate to support an animal while it runs.

One reason that muscle work may not be a major determinant of energy expenditure during steady level locomotion is that much of the energy that might otherwise be lost is recovered by efficient PE and KE exchange during walking, or is effectively recovered via elastic strain energy in tendons and ligaments (see following section). In addition, many limb muscles may contract with little length change and hence do little or no net work. Instead, these muscles may undergo a brief period of stretch or isometric force development which facilitates more economical force development than when a muscle shortens as it develops force. This is because a muscle generates less force when it shortens more rapidly (see Chapter 2, section 2.1.3). When muscles contract isometrically, or when they are briefly lengthened, they can generate greater force. Under these conditions, a smaller fraction of the muscle's fibers must be activated to generate a given force. This increases a muscle's 'force economy' by decreasing the amount of energy (ATP) that must be expended per unit force produced.

As we discussed in Chapter 2 (section 2.1.9 and Fig. 2.11), the architecture of a muscle is also important in determining a muscle's force economy. Shorter pinnate-fibered muscles are more economical in their design for generating force than longer parallel-fibered muscles. Short-fibered muscles also often attach to long tendons which favor increased elastic energy savings (see below). The importance of force economy was first demonstrated in the lateral gastrocnemius of running turkeys by Roberts *et al.* (1997) and has also been shown in the leg muscles of hopping wallabies (Biewener *et al.* 1998a). In these muscles, force is developed rapidly under near isometric conditions when the animal's limb lands on the ground (Fig. 3.13). In some instances, a brief initial stretch of the muscle's fibers may allow the muscle to generate 1.5–1.8 times greater force than when isometric, providing an additional 50–80 per cent energy savings in terms of force economy. In both running turkeys and hopping wallabies, the leg muscles do little work. Instead, elastic energy savings in the tendon represents 60–96 per cent of the total work done by the muscle–tendon unit as a whole.

3.12   Tendon springs and muscle dampers

The long tendons and foot ligaments of many vertebrate runners (Fig. 3.14), including those of humans, are important sources for storing and recovering elastic energy which minimizes the work that the limb muscles must do to restore an animal's PE and KE while running, hopping or galloping. Measurements of the properties of vertebrate tendons (mainly for various mammals and birds) show that, when stretched, tendon provides a highly resilient elastic tissue (Fig. 3.15), returning up to 93 per cent of the energy that is stored when it is stretched. The slight loss of energy (7 per cent), reflected by the 'hysteresis' in the loading versus unloading behavior of tendon, is not uncommon for many tissues and reflects the fact that tendons display both viscous (energy lost) and elastic (energy recovered) properties. Hence such

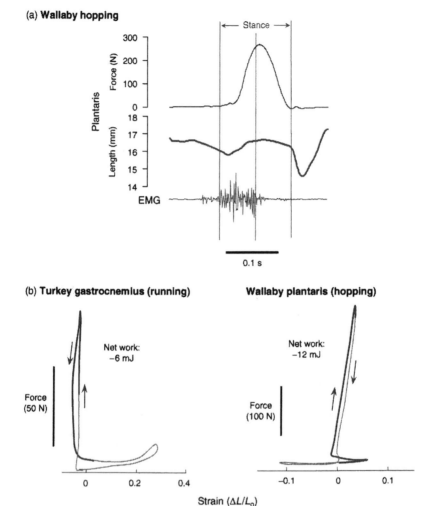

Fig. 3.13 *In vivo* force and length change patterns of muscles which do little net work but favor force economy and tendon elastic energy savings. (a) Recordings from the plantaris hind leg muscle of a hopping wallaby. (b) Graphs of force versus fractional length change (strain) of the gastrocnemius of a running turkey and the plantaris of a hopping wallaby. The darker portion of each graph denotes the time during which the muscle is activated based on its EMG. Arrows indicate the direction of force versus length change over one locomotor cycle. (Adapted from Roberts *et al.* (1997) and Biewener *et al.* (1998).)

tissues are referred to as being 'viscoelastic'. In the case of tendon, its elastic properties dominate its overall behavior. In addition to losing some energy when unloaded, tendon also exhibits 'non-linear' stress–strain behavior, having a characteristic J-shaped curve. At low stresses the 'toe' region of the curve

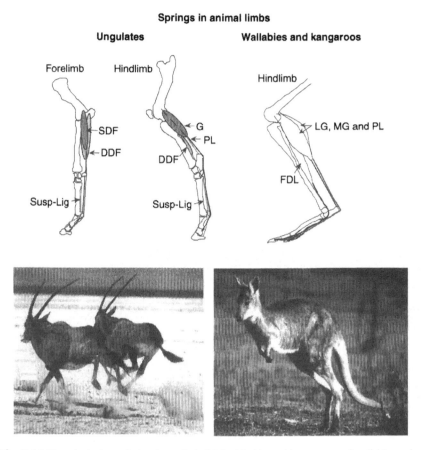

Fig. 3.14 The principal muscle–tendon units in (a) the hind legs of kangaroos and wallabies and (b) the limbs of horses and other ungulates that are specialized for economical force generation by short-fibered muscles which attach to long tendons for elastic energy savings.

has a lower slope which increases with increasing stress, becoming relatively constant at moderate to higher stresses (and strains). This indicates that the tendon is less stiff at lower stress than when it operates at moderate to high stresses (this is believed to result from the relative sliding and re-alignment of collagen fibrils within the tendon at low stresses, which are then pulled on directly when the tendon's stiffness increases). For most tendons and ligaments that have been tested the toe region is fairly small, and above a strain of about 2–3 per cent the slope of the tendon's stress–strain curve can be considered nearly constant. Over this linear range of behavior, tendon has an elastic modulus about 1.2 GPa. Tendon ruptures at stresses in the range 100–120 MPa, indicating a failure strain of about 8–10 per cent. From measurements of the stresses that various mammalian tendons experience when an animal runs, the

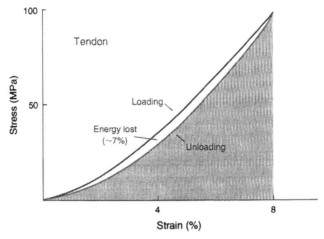

Fig. 3.15 Stress–strain curve of vertebrate tendon showing the hysteresis that exists between the loading and unloading phases of the cycle. The area within the hysteresis loop reflects the energy that is lost as heat due to the viscous properties of the tendon. For vertebrate tendons this is generally as low as 7 per cent. The shaded area beneath the unloading curve represents the energy that is recovered elastically (93 per cent of the total energy that stretched the tendon) and can be used to offset muscle work. Although tendon is less stiff at lower stresses, its overall properties are well suited to serve as an effective spring for energy recovery. Most tendons operate at strains up to about 5–6 per cent.

normal range of tendon strain appears to be limited to a maximum of about 5–6 per cent (Bennett *et al.* 1986).

The non-linear and slightly viscoelastic behavior of tendon contrasts with the linearly elastic behavior that was used to provide a simple overview and comparison of the properties of various biological materials introduced in Chapter 1. Nevertheless, the fact that tendon is stiff (1.2 GPa) and highly resilient (93 per cent) over much of its functional range makes it an excellent material for storing and recovering elastic strain energy to minimize the mechanical work and metabolic cost of locomotion. A reasonable estimate for the strain energy recovery of vertebrate tendons is

$$0.5F\Delta L \times 0.93 = 0.465F\Delta L$$

where F is the maximum force transmitted by the tendon and ΔL is its total change in length. In wallabies and kangaroos, elastic energy recovery is substantial enough that when these animals hop at faster speeds their metabolic rate of energy use does not increase (see Chapter 9, Fig. 9.2). This remarkable observation is in contrast with all other terrestrial animals that have been studied, even those with highly specialized tendons such as horses and antelope. It has been estimated that tammar wallabies hopping at 6 m/s reduce their metabolic energy expenditure by 50 per cent through elastic energy recovery in their leg tendons alone (Biewener and Baudinette 1995). Additionally, female tammar wallabies can carry pouch young weighing up to 15 per cent of their own weight for free (Baudinette and Biewener 1998); the additional cost that would otherwise be incurred due to the joey's weight is conserved by increased elastic energy savings in the mother's leg tendons. The ability of these animals to carry loads for free is another unique feature of their hopping compared with other vertebrate runners that have been studied. However, the ability to conserve elastic energy by hopping is not universal. Small heteromyid rodents appear to have tendons that are too thick to enable them to develop sufficient levels of strain energy for the forces that are required during steady speed hopping (Biewener *et al.* 1981). Instead, their relatively thick leg tendons appear to be better designed for jumping and predator escape (Biewener and Blickhan 1988). The inability of small hopping rodents to store significant elastic energy in their tendons appears to result mainly from their size. If scaled up (isometrically) to the size of a much larger kangaroo or wallaby, a kangaroo rat would be well-suited for effective elastic energy savings.

Although other vertebrates do not appear to achieve the same degree of metabolic energy savings as wallabies and kangaroos, tendon energy savings are clearly important to their ability to conserve energy that their muscles would otherwise have to perform as mechanical work. Elastic energy savings in the Achilles tendon and ligaments of the foot are estimated to reduce muscle work by up to 50 per cent in a human runner (Ker *et al.* 1987); in horses, elastic energy savings are estimated to reduce muscle work by up to 40 per cent at a trot and a gallop (Biewener 1998). Measurements in running turkeys indicate that elastic energy storage can be quite significant, and indirect estimates also indicate the importance of tendon elastic energy savings in dogs and other trotting animals (Cavagna *et al.* 1977).

In most cases, elastic energy savings are not free. Metabolic energy must be consumed by the muscles to generate the forces needed to operate tendon springs, but as we discussed above this can be accomplished at lower cost when a muscle contracts isometrically or is briefly stretched. Energy cost is also reduced by muscles that have shorter and typically more pinnate fibers, since this reduces the volume of muscle that must be recruited to generate a given force (see Chapter 2, section 2.1.9). This raises the question of why have a muscle that consumes energy at all? The reason very likely is that, muscles and their tendons must not only transmit force effectively, with the possibility of enhancing tendon elastic energy savings, but they also must provide control of movement (see Chapter 10). Without an active force-generating muscle attached to a tendon, the control of length change and displacement of a limb segment becomes entirely passive. While this may optimize energy savings by the tendon (or ligament), by eliminating the energy-consuming muscle control of length is also lost.

In addition to control of length, one other role appears to be important for retaining an energy-consuming muscle that transmits force to a passive tendon. Some of the most highly specialized muscle–tendon systems for elastic savings are found in the legs of horses and other ungulates (Fig. 3.13). Many of the muscles that attach to the distal leg tendons of these animals have extremely short fibers. In some muscles the fibers are as short as 5–7 mm and attach to tendons that are nearly 1 m long. Until recently, the role of these short fibers has been a mystery. Their contraction cannot provide any useful control of distal tendon length because the stretch of the tendon (30 mm at 3 per cent strain) greatly exceeds the length of the fibers themselves. Because of this, these short-fibered muscles had previously been viewed as a vestige of evolution, dating back to when horses and their ungulate relatives were much smaller and less cursorial. However, in recent experiments Wilson *et al.* (2001) have shown that mechanical vibrations produced by the impact of a horse limb with the ground are effectively damped out by these short-fibered muscles. Indeed, damping was much more effective when the muscles were actively stimulated than when they were passive. Such vibrations might otherwise have damaging effects on the animal's joints and tissues over a longer period of use. The importance of short-fibered muscles as dampers, or shock absorbers, explains their retention in modern ungulates and very likely their role in other animals as well.

3.13 Summary

Terrestrial locomotion encompasses an amazing array of legged animals and yet basic principles of locomotor mechanics and energetics are found across this diversity. These patterns emerge as animals of differing size and construction must contend with support against gravity when moving on land. Walking gaits involve pendular exchange of the body's potential and kinetic energy to reduce muscle work and energy cost. Running gaits involve an elastic bounce of the body over the supporting limb, allowing for elastic energy recoil of spring elements in the limb to reduce muscle work. In hopping kangaroos and wallabies, this energy recovery can be quite remarkable, and in humans it can also provide 50 per cent of the work that leg muscles would otherwise have to perform. Reduced energy cost is also favored by muscles that generate force economically. This is best achieved by having shorter-fibered muscles which generate forces under more isometric conditions. In addition to having an important effect on length control and energy use, muscle architecture also plays an important role in damping unwanted vibrations of the limb. These patterns of whole-body energy exchange, and the roles of the muscles and tendons underlying them, apply to two-legged, four-legged, six-legged and even sideways eight-legged runners! Legged locomotion has the distinct advantage of providing an effective means of transport over uneven and unpredictable terrain.

In order to move faster, animals change gait and reduce the time that their limbs remain in contact with the ground, requiring them to be dynamically stable. Basic mechanical properties of limbs and joints (stiffness, energy absorption, spring energy recovery and work output) provide dynamic stabilization and may greatly simplify the motor control task of the nervous system. Running faster also requires effective support of greater forces. In animals of very different size, but built of similar materials, adjustments in the organization and posture of the limb are important for regulating the level of force and stress that must be transmitted. Adjustments of limb posture and body size are also important influences on the ability of an animal to maneuver and turn. Before considering the energetic implications of terrestrial gaits and mechanisms, we shall next dive into swimming and then come up for air to discuss flight, which both involve movement through fluids. While gravity remains of concern for flying animals, it can largely be ignored by those that swim.

4 | Movement in water

Swimming animals span an enormous range in size and shape. The largest living organisms on Earth, the whales, are 10^{10} greater in mass than swimming bacteria, yet both must contend with the same fluid environment. In spanning such a broad size range, aquatic organisms have evolved a variety of propulsive mechanisms to move through the water. These mechanisms range from the oscillatory movements of the fins and body axis of fish and the flukes of whales to the motions of bacterial flagella. They also include the jetting of squid and the rowing of cilia. In this chapter, we shall seek to identify common principles of design which underlie these various propulsive mechanisms that animals use and consider the consequences of size on these mechanisms. As for other modes of locomotion, differences in size play an important role in dictating which physical properties of the medium are most important to propulsive mechanisms in a fluid. Finally, the energetic requirements of the various modes of aquatic locomotion will be considered in relation to terrestrial and aerial locomotion.

4.1 Thrust and drag

As in all other environments, locomotion in water involves the use of body appendages or body surfaces to generate propulsion, or thrust, by pushing against the surrounding medium. In the aquatic environment, this involves the active transfer of momentum from moving portions of an animal's body to the water surrounding it. The momentum (mass × velocity) that is transferred can be thought of as the mass of water that is accelerated by the animal's body to a given average velocity. The rate at which the animal transfers momentum to the water (i.e. mass × acceleration) determines the amount of **thrust** that it generates:

$$T = mv/t \qquad (4.1)$$

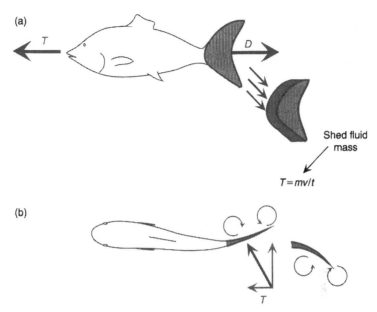

Fig. 4.1 (a) Diagram showing thrust and drag forces acting on a swimming fish and the momentum that it transfers to the fluid by its caudal fin. (b) Dorsal view of the fish showing the net propulsive force acting on the fish with the anterior component of thrust T that the tail produces. The shed vortices of fluid are depicted for both right and left movements of the tail.

where v is the velocity of fluid of mass m moved per unit time. Thus, one way for a fish to generate more thrust and swim faster is to beat its tail at a higher frequency (increasing v/t). Thrust is defined here as the force exerted by the fluid on the animal's body *in reaction to* the fluid being accelerated by the animal's body (Fig. 4.1), and generally acts in the direction of an animal's motion through the fluid.

At the same time that an animal must generate thrust to move forward, it is resisted by the movement of the fluid past its own body. The resistive force exerted by the fluid on its body is termed **drag**. Drag acts opposite to an animal's forward motion and hence opposes thrust (Fig. 4.1). As a resistive force, drag represents the rate at which momentum is lost by the animal to the fluid moving by its body. Therefore effective swimming requires propulsive mechanisms that enhance thrust and reduce drag. Before discussing these mechanisms, we need to consider the underlying hydrodynamic basis of drag.

4.2 Inertia, viscosity and Reynolds number

When a swimming animal moves through the water its motion is enhanced by its own inertia and is resisted by the fluid. As noted above, the inertial forces required to keep an animal's body moving depend on its mass and changes in its forward velocity. Resistive forces due to fluid movement past the animal's body depend on the 'stickiness', or viscosity, of the fluid and the pressure exerted by the fluid on the organism. These two components of drag are often referred to as 'friction drag' and 'pressure drag'. Their relative importance also depends on the size and speed of the animal. Pressure drag is most important for larger animals that swim at faster speeds, whereas viscous drag is most critical at small size and slow speeds. As will be explained below, pressure drag results from the pressure gradient developed from the front to the back of the swimming organism due to flow separation. Because it is not a hydrostatic pressure it does not change with swimming depth. Viscosity represents a measure of the resistance of a fluid to being sheared, or more precisely to its rate of shear. Shear represents the relative deformation or sliding of parallel layers of a fluid (or a solid) with respect to one another. The stickier the fluid, the more the fluid will resist being sheared and hence the greater its viscosity. Because viscosity μ depends on the ratio of shear stress to shear rate, it has units of stress $\times$ time (Pa s). Mineral oil (glycerin) has a viscosity of 1.49 Pa s at 20 °C, which is 1.49×10^3 times greater than the viscosity of fresh water (0.001 Pa s at 20 °C). Differences in the viscosity of these, and other, fluids become useful in practice when constructing mechanical models of organisms in order to study their hydrodynamic performance under simulated biological conditions. In terms of understanding the fluid dynamics of aquatic locomotion, however, the viscosity of the aquatic environments inhabited by biological organisms can be considered to be essentially the same (seawater has a viscosity that is generally only about 7 per cent greater than fresh water).

The relative importance of inertial forces relative to viscous forces during locomotion through a fluid is defined by the Reynolds number (Re), a dimensionless parameter that is central to the dynamics of flow:

$$Re = \rho l v / \mu \qquad (4.2)$$

where ρ is the density of the fluid, l is a characteristic length (body length, fin length, or wing length in the case of a flying animal), v is the organism's (or appendage's) forward velocity relative to the fluid and μ is the viscosity of the fluid. Although not derived here, Re can be shown to be the ratio of inertial ($\rho S v^2$) to viscous ($\mu S v / l$) forces experienced by an organism (where S is some measure of the organism's surface area exposed to the flow). Vogel (1994) gives an easily readable and entertaining discussion of the physical basis for Re and its importance to biological fluid mechanics. The key point is that Re provides a metric for judging the relative importance of inertial to viscous forces affecting the organism's movement through a fluid medium. At equal Reynolds numbers, flow characteristics are the same. This allows one to model

Table 4.1

Organism or Aircraft	Reynolds Number
Concorde flying at 600 ms^{-1}	30,000,000,000
Mini-light aircraft flying at 50 ms^{-1}	10,000,000
Tuna swimming at 3 ms^{-1}	10,000,000
Duck flying at 20 ms^{-1}	300,000
Dragonfly moving at 7 ms^{-1}	30,000
Trout fry swimming at 0.2 ms^{-1}	3,000
Small butterfly moving at 1 ms^{-1}	3,000
Copepod burst swimming at 0.2 ms^{-1}	300
Flapping wings of a fruit fly moving at 0.2 ms^{-1}	30
Sperm swimming to advance the species at 0.2 mms^{-1}	0.03
Bacterium swimming at 0.01 mms^{-1}	0.00001

Adapted from Vogel (1994).

flow conditions at different scales (e.g. to build a small model of a very large organism, or vice versa) by keeping Re the same. It is central to the study of swimming and flight, as both depend on the same fluid dynamic principles. Consequently, we shall return to Re when discussing flight in the next chapter.

At high Re (>100) inertial forces dominate, but at low Re ($\ll$1) viscous forces reign. This has important consequences for the propulsive mechanisms and design for locomotion at high and low Re. The Reynolds numbers at which various aquatic and flying organisms operate are given in Table 4.1. In contrast with a 3-m tuna which glides several body lengths through the water when it stops swimming because of its own inertia, a 200-μm ciliate stops almost immediately once it ceases to swim. Berg (1983) provides a dramatic example of the absence of inertia at very low Reynolds numbers, calculating that a bacterium coasts 0.1 Å, or the diameter of a hydrogen atom, once its flagellum stops beating! At these low Reynolds numbers (typically at extremely small body sizes), aquatic locomotion becomes counter-intuitive with respect to our own experience as large-bodied swimmers (we operate at Re $\approx 10^6$).

In the intermediate range of Re, between about 0.1 and 100, both inertial and viscous forces operate and change as a function of Re. This is a messy range for which theory does not work very well. One means of getting around theory is to measure the drag force on an organism, or a model of an organism, in a flow tank (see Fig. 4.7) and relate this to its shape and velocity in order to calculate a drag coefficient C_d. The drag coefficient essentially represents an experimental measure of the ratio of measured drag force D to the theoretically predicted drag force:

$$C_d = 2D/\rho Sv^2 \tag{4.3}$$

or

$$D = C_d \rho Sv^2/2. \tag{4.4}$$

Consequently, the drag force experienced by an organism depends on the drag coefficient (measured for a given Re), the density of the fluid ρ, some measure of the surface area S of the organism (for swimming animals this is most often the frontal area projected to the oncoming flow) and the square of the animal's forward velocity. As we know from riding a bicycle and concerns of automobile fuel economy, drag forces depend most heavily on velocity ($\propto v^2$), However, at low Re (low v) the drag coefficient is large, reflecting the importance of viscous forces, whereas at higher Re (high v) the drag coefficient decreases as inertial forces become more important. Because drag also depends on the fluid density, it exerts a much larger force at a given speed in water than in air. Consequently, swimming animals typically encounter much higher levels of drag and move at much slower speeds than flying animals. Finally, except at extremely small size, shape is important to determining the magnitude of drag. Therefore streamlined shapes which reduce drag by reducing the amount of energy lost to the wake (described in more detail below) are favored over blunt or irregular shapes.

4.3 Steady flow: drag and streamlines

The physical basis for viscosity and drag is best seen by considering streamlines of a fluid's motion under conditions of steady flow past a solid object (Fig. 4.2). Streamlines can be thought of as representing the paths of movement of individual fluid particles at different locations within a field of flow. In practice, streamlines can be visualized within a flow tank (see Fig. 4.7(c) below) by adding dyes to the fluid at discrete locations or by mixing small neutrally buoyant particles to the fluid and observing their motion on video or film. An important principle underlying fluid mechanics is 'continuity of flow', which requires that the volume flow rate of fluid moving past an organism is constant. In other words, all fluid must be accounted for—akin to the conservation of energy principle, fluid can neither be created nor lost. In an idealized fluid (zero viscosity) (Fig. 4.2(a)) the streamlines move symmetrically past the long cylindrical object. These represent the theoretical streamlines for **laminar** flow around the object, which means that the streamlines largely remain parallel to one another. Analogous to topographic maps, streamlines that are closer together represent increased velocity of flow. In this example flow is greatest lateral to the object and has local zero-velocity regions, or stagnation points, at the front and rear of the object. Because of the symmetry of flow, the pressures exerted on the object balance out and suggest that the drag on object should be zero.

However, in practice this is never the case. Symmetry is not achieved for two reasons. First, real fluids have viscosity, so that all water in contact with the surface of the organism is by definition stationary and therefore has a local zero velocity. This is often referred to as the 'no-slip condition'. Further away

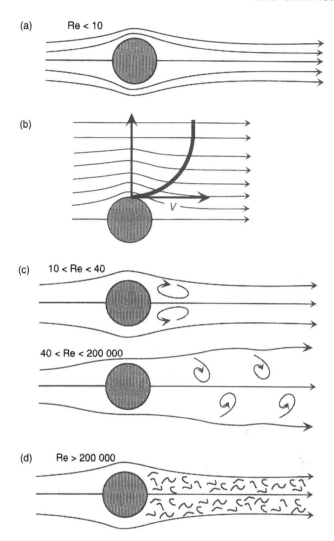

Fig. 4.2 (a) Idealized streamlines of fluid moving past a cylinder perpendicular to the flow. (b) The (parabolic) gradient of flow moving away from surface of an object produced by the shear imposed by fluid drag. (c) Flow at low to moderate Re showing stationary and shed vortices. (d) Flow at high Re showing a turbulent wake (Adapted from Vogel (1994) with permission from Princeton University Press). See text for further details.

from the organism's surface, the velocity of fluid movement increases parabolically up to the free-stream velocity of the fluid moving past the organism (Fig. 4.2(b)). For an organism swimming through stationary fluid the free-stream velocity is equal but opposite to the organism's forward swimming velocity. This velocity gradient represents skin friction drag, which depends on the viscous interaction of fluid layers that are sheared as they move over the surface of the organism. Skin friction drag increases in proportion to the surface area of the organism exposed to flow, and this causes a net deceleration in the flow of fluid past the organism.

Flow asymmetry is also produced by pressure drag, which develops because in real situations the dynamic pressure exerted by the water moving past the front of the organism is greater than the pressure developed at its back. The reason for this difference in pressure is that the energy transferred to the water when it is accelerated as it passes around the sides of the organism is not completely returned to the organism as the water moves along its downstream end. This loss of energy occurs due to flow separation along or downstream of the organism. When flow separation occurs at low to moderate Re, attached or shed vortices develop which dissipate the energy of the fluid as it moves past the organism (Fig. 4.2(c)). At faster flows with higher Re the vortices break down and the flow becomes **turbulent** (Fig. 4.2(d)). Turbulent flow behind the organism causes an even greater loss of fluid energy. The loss of energy results in a decrease in the pressure on the downstream side of the organism, making the dynamic pressure difference between the upstream and downstream sides greater, which increases the pressure drag experienced by the organism.

4.3.1 Steady versus unsteady flow

Before considering specific examples of aquatic locomotion, one more issue needs to be addressed. This is the matter of steady versus unsteady flow. Almost all our discussions so far have assumed a steady or constant state of flow past the organism. However, in most biological cases, propulsion by reciprocating appendages or an undulating body necessarily involves unsteady flow (Daniel 1984; Dickinson 1996). This is because the movements of the body and/or the appendages must be accelerated from rest to generate propulsive force and must then be decelerated before initiating a subsequent propulsive stroke. Consequently, in most cases fluid propulsion involves not only overcoming the force due to drag, which depends on the steady speed of the animal, but also responding to forces brought about by changes in the velocity of flow. Changes in the velocity of flow introduce what is known as an 'acceleration reaction' force that adds to (or may subtract from) the drag force acting on the animal (Daniel 1984). When any part of an animal's body accelerates through a fluid, it entrains and must also accelerate an additional mass of fluid. The acceleration reaction force F_{ar} depends not only on the mass of the animal's own body that is being accelerated (ma), but also on the 'added mass' of the fluid that is being accelerated as well:

$$F_{ar} = ma + C_a \rho V a = a(m + C_a \rho V) \tag{4.5}$$

where V is the volume of fluid that is being accelerated, ρ is its density and C_a is the 'added mass coefficient'. The added mass $C_a \rho V$ depends on the shape of the propulsive element and the volume of water that it causes to be accelerated. Similar to the drag coefficient, the added mass coefficient is lower for more streamlined bodies. The acceleration reaction is likely to be most important at slower speeds or when an animal accelerates from rest. Despite its potential importance, few measurements of acceleration reaction relative to drag force have yet been made. Because of the unsteady nature of flow in most biological situations, it is a matter that deserves further study.

4.4 Swimming fish, mammals and cephalopods: movement at high Reynolds number

Aquatic locomotion at high Re ($\gg$100) is dominated by inertial forces. Viscous forces become relatively unimportant in the Re regime at which many fish, humans and cetaceans operate (10^5-10^8). Even smaller invertebrate swimmers (e.g. crustaceans, copepods and water beetles) in the Re range from 300 to 10 000 rely largely on inertial mechanisms for propulsion. At high Re, drag is dominated by pressure drag, rather than skin friction drag. Therefore hydrodynamic efficiency is enhanced by keeping the drag coefficient (eqn (4.3)) as small as possible. This is best achieved by having a streamlined body shape (Fig. 4.3(a)). The reason that a tuna would coast several body lengths if it stopped swimming is largely due to its streamlined body shape. In contrast,

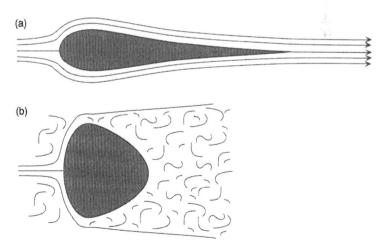

Fig. 4.3 Effect of (a) a streamlined body shape and (b) a bluff body on flow and turbulence. See text for further details.

streamlining is unimportant in the low Re regime of a swimming bacterium (see section 4.6).

Streamlining enhances laminar flow past the body. Streamlined bodies have a bluff shape in front but a gradual taper to the rear. The taper is the trick. With a gradual taper, flow is less likely to separate and become turbulent as it moves by the body. Recall that flow separation and the resultant creation of a large and turbulent wake is the primary basis of energy loss and drag at moderate to high Re. Flow separation is more likely to develop the faster the flow and/or the more bluff the body's shape (Fig. 4.3(b)). Having a tapered body allows the fluid to decelerate gradually as it moves back along the animal, favoring the maintenance of laminar flow. As a result, the body is essentially pushed along by the wedge-like closure of fluid behind it. In contrast, bluff bodies result in abrupt separation and rapid fluid deceleration, which create a turbulent wake. This causes a loss of pressure behind the body and leads to a large pressure drag. Animals that swim or jet over a broad range of moderate to high Re, such as fish, whales, dolphins, squid and crayfish, generally have streamlined body shapes. Although streamlining incurs greater skin friction drag (due to the increase in surface area associated with a long tapered length of the body), this is largely unimportant at high Re and is greatly offset by the reduction in pressure drag.

At moderate to high Re, the general mechanism for thrust production is to accelerate a mass of fluid backward so that a net reaction thrust is exerted on the animal to propel it forward. By accelerating fluid backward the animal's muscles generate thrust by transferring momentum to the fluid. This is the case whether thrust is produced along most of the animal's body axis (e.g. eel), concentrated at a caudal fin (e.g. tuna), a pectoral fin (e.g. stingray) or a fluke (e.g. killer whale), or achieved by ejecting a bolus of fluid from within the animal's body (e.g. squid). These different styles typically emphasize differing aspects of swimming performance. Whereas pectoral fin propulsion in fish is more commonly associated with slower, more maneuverable swimming and depends on *drag-based* propulsion, caudal fin propulsion is typically associated with faster open-water swimming which depends on *lift-based* propulsion. These propulsive mechanisms and the effects of body shape on swimming style are discussed below.

4.4.1 Undulatory swimming

Undulatory or **anguilliform** swimmers typically have relatively elongate bodies. Thrust along the undulating body of a fish, salamander, or sea snake can be considered in the context of the thrust that a local segment of the body produces (Fig. 4.4). At high Re, thrust is dependent on the rate at which momentum is transferred, or shed, to the fluid. This can be summed over the length of

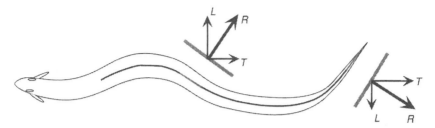

Fig. 4.4 Thrust in an undulatory swimmer. Thrust *T* is produced locally by exerting a rearward component of the reaction force *R* on the fluid adjacent to the body surface. This results from the angled orientation of the body which is achieved by the sinusoidal waves of bending that travel down the animal's body. This produces a laterally directed force *L* as well as a component of thrust. The lateral forces are cancelled out over time as the animal's body bends back and forth. These forces are summed along some length of the animal's body. Undulatory swimmers typically generate most of their useful thrust with the posterior half of their body.

the animal to calculate the total average thrust that a steadily undulating body produces. Undulatory swimming involves sinusoidal waves of body bending which travel down the body axis. Anguilliform swimmers tend to pass waves along much of their body but, as with other undulatory swimmers, the amplitude of the wave increases as it passes to the tail (Gillis, 1996). As any segment of the body becomes incorporated into the backward-traveling wave, it forms an angle to the direction of the animal's forward movement and produces a local force on the animal's body that has two components (lateral and backward). The lateral force is cancelled out by symmetric lateral oscillations of each segment as the body bends in opposite directions. The backward forces produced by these undulating body segments are summed to generate a net propulsive force (i.e. thrust) which accelerates the animal forward. When the net thrust force is balanced by the resistive drag force, the animal swims at a constant forward speed.

Most fish, such as cod, perch and trout, exhibit a more general style termed **carangiform** swimming, in which the posterior third of the body undulates, with the amplitude of undulation increasing toward the tail where it is maximal. These fish are also distinguished from anguilliform swimmers in that they are shorter with respect to the length of the bending wave that passes backwards. (Although, by definition carangiform fish have body lengths that are less than one wavelength of the traveling bending wave, fish body shapes and swimming styles span a continuum from eel-like fish which utilize a large majority of their body length to generate thrust to fish which increasingly concentrate thrust at their caudal fin.) In order to enhance thrust, the surface area of the caudal fin is enlarged. Because momentum is conserved, thrust production of the swimming animal can be most simply analyzed as balancing the momentum mv of the animal relative to that transmitted to the water as thrust:

$$mv + \int F \mathrm{d}t = -MV \qquad (4.6)$$

where M and V are the mass and velocity of the water moving in the opposite direction of the animal's travel. This equation is an oversimplification of most biological situations because of the unsteady nature of flow, in which the velocity of the water is not constant but changes over time as it is accelerated and its direction is changed by the moving fin (Webb 1982 and see section 4.3.1). Nevertheless, it serves to illustrate two useful points. First, by enlarging the surface area of the fin, more water can be accelerated to generate thrust and thus overcome drag. Second, the drag on the tail's movement can be reduced by the tail not having to oscillate at as high a velocity as would be required if it were smaller.

4.4.2 Caudal fin or fluke swimming

At the opposite extreme of swimming style, **thunniform** swimmers and cetaceans tend to avoid undulatory movement over almost all of their body length (in order to reduce drag) by emphasizing lateral undulation of their caudal fin or flukes. The lunate shapes of the fins and flukes of these animals (Fig. 4.5) and their similar swimming modes represent one of the striking examples of convergent evolution. Because aquatic mammals must contend with the same physical forces for effective fluid propulsion as a large swimming fish, whales and dolphins have evolved caudal flukes with a lunate shape, analogous to the caudal fins of thunniform fish, but which they oscillate in the dorsoventral plane (Fish 1996). In contrast with the drag-based propulsion

Fig. 4.5 Swimming animals, such as tuna and whales, are able to achieve lift-based thrust. This is similar to the lift that flying animals achieve with their wings. The tuna's caudal fin or the whale's flukes generate a lift force (L) which, by being inclined in the direction of swimming, has an anterior component of thrust (T). This balances the total drag force (D) on the fish, whereas the local drag experienced by the tail is represented by 'd'. Lateral or dorsoventral forces are cancelled out by repeated beating of the tail or flukes. See text for further details.

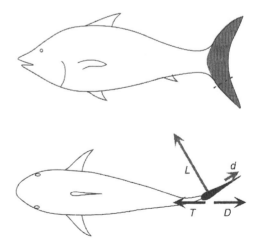

of undulatory swimmers, these animals are believed to produce thrust by a lift-based mechanism. Lift is explored more extensively in Chapter 5 when discussing flight, but its introduction is helpful here as it provides an important mechanism for producing thrust in these fast-swimming animals.

When viewed in cross-section, the fins of tuna or flukes of dolphins have a streamlined shape (Fig. 4.5). As such, they can be considered **hydrofoils**. By oscillating back and forth, they encounter the oncoming water at a shallow angle, referred to as the 'angle of attack'. This induces an asymmetric flow of water over them resulting in a velocity differential on either side of the hydrofoil (the velocity of flow is greater on the side of the hydrofoil with the greatest distance between the stagnation points of flow). This velocity differential can be considered to represent the sum of a translational velocity component and a circular component (see Chapter 5, Fig. 5.2). Fluid does not actually flow around the hydrofoil (except when shed at its tip), but the effective circulation of flow (i.e. its **vorticity**) that is established each time the fin beats to either side produces a net force perpendicular to the direction of flow over the fin. This force is termed **lift** (for further explanation of lift see Chapter 5, section 5.1). Lift can also be thought of as resulting from the pressure differential that is established by the difference in velocity on each side of the hydrofoil. Based on Bernoulli's principle (pressure varies inversely with the velocity of flow), water moving at a slower velocity (upper side of the hydrofoil in Fig. 4.5, or the fish's right side) exerts a greater pressure on the hydrofoil than water moving at a faster velocity (on its left side). Hence, a net lift force L is generated in the direction of lower pressure. By being angled to the direction of oncoming water, the lift produced by the hydrofoil has an anterior component. With the appropriate angle of attack significant lift is achieved relative to drag D, generating a net anteriorly directed thrust force T. Lateral components of hydrodynamic force in the case of fins or dorsoventral components in the case of flukes cancel out over successive tail beats. In contrast with dolphins and whales, phocid seals and walruses use pelvic oscillations to undulate paired hindflippers in the lateral plane, whereas fur seals and sea lions use pectoral (fore) flippers as oscillatory hydrofoils (Fig. 4.6). Both of these flipper-based propulsive mechanisms also achieve lift-based thrust (Fish 1996).

The improved efficiency of lift-based propulsion at faster speeds is similar to that of airfoils (wings of insects and birds) in flight, resulting from increased airflow induced by the animal's own forward movement in addition to that of the movement of the fin with respect to the animal.

4.4.3 Tail shape: homocercal versus heterocercal tails

Tail shape is an important feature of fish, influencing the way that the tail works and how it generates thrust. As noted above, fish which swim at high speeds for long distances, such as tuna and mackerel, have symmetrical lunate-shaped

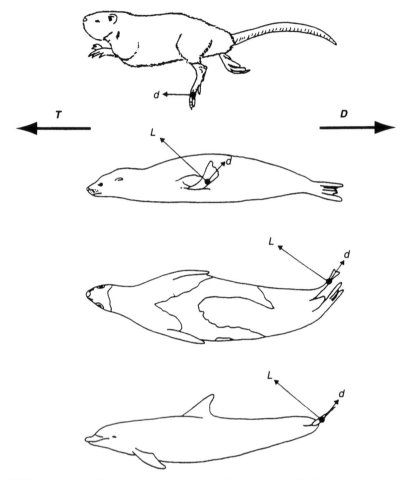

Fig. 4.6 Various mammals which have evolved aquatic locomotor specializations use different structural appendages and propulsive mechanisms (drag and lift based) to propel themselves through the water. See text for further details. (Adapted from Fish (1996), with permission from Allen Press.) T: anterior component of lift force (C) that overcomes total drag force (D). d is the local drag force on the appendage serving as a hydrofoil.

tails which generate thrust through lift-based propulsion. In fact, most bony fish possess symmetrically shaped tails that are described as being **homocercal**. Many of these fish have shorter more stocky tails which generate thrust by means of drag-based propulsion. Because of their symmetry, homocercal tails are thought to generate thrust in line with the animal's movement. Recent kinematic analysis and flow visualization (Gibb et al. 1999) indicate that, in at least some species with homocercal tails, differences in stiffness through the depth of the tail (from top to bottom) result in non-uniform movement of the caudal fin as it beats back and forth. This suggests that the net force produced by the tail may not be simply to generate anterior thrust.

In contrast with most fish that have homocercal tails, sharks and some other fish (e.g. sturgeon) have asymmetrical or **heterocercal** tails (Fig. 4.7). Whereas homocercal tails are thought to provide uniform forward thrust, heterocercal tails are believed to produce lift in addition to thrust. This is because the ventral lobe of the tail is reduced relative to the extended and stiffened dorsal portion of the tail. Recent three-dimensional kinematics and flow visualization of tail and

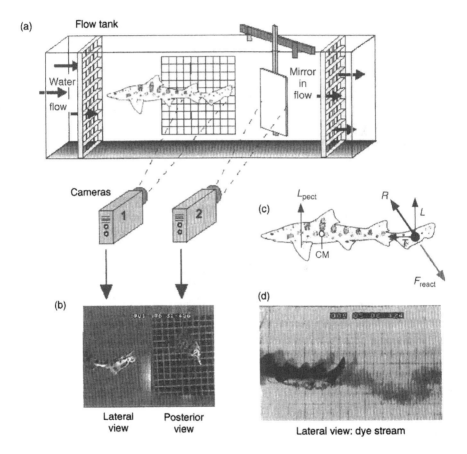

Fig. 4.7 (a) A flow tank is used to study swimming in many fish, including the leopard shark shown here. (b) Using two video cameras and a mirror behind the animal, both lateral and posterior views of the tail's motion can be seen. (c) The heterocercal tails of sharks produce a net upward force L which must be balanced by the lift L_{pect} produced by the pectoral fins to prevent the shark from pitching about its center of mass (CM, open circle). (d) Dye streams showing the pattern of flow produced by the heterocercal tail of the shark confirm this. (Adapted from Ferry and Lauder (1996), with permission from the Company of Biologists Ltd.)

water movement (Ferry and Lauder 1996) (Fig. 4.7) has confirmed this function of heterocercal tails for swimming sharks, but subsequent work indicates that lift production by the heterocercal tail of sturgeon is less significant. The lift produced by the heterocercal tail of sharks helps to counteract the negative buoyancy of these animals, but also generates a pitching moment about the shark's center of mass. This is resisted by additional lift produced by the shark's pectoral fins. Therefore lift production and the presence of a heterocercal tail are central to the body form and swimming function of many cartilaginous fish, all of which are negatively buoyant despite the retention of body oils to reduce their negative buoyancy. Bony (teleost) fish which possess swim bladders, on the other hand, can achieve neutral buoyancy and, because of this, do not require constant swimming or an asymmetrical tail shape to generate lift in order to maintain their position in the water column.

The evolution of a gas swim bladder has enabled a wide range of swimming morphologies and behaviors within teleost fish and probably underlies much of their success as the most speciose class of vertebrates. Indeed, their ability to be neutrally buoyant has allowed them to become exceptionally maneuverable swimmers, and this has led to the use of the pectoral fins as flexible propulsive organs.

4.4.4 Pectoral fin swimming

Paired pectoral fin locomotion probably involves both drag- and lift-based mechanisms of propulsion. These mechanisms are the result of oscillatory movements of the pectoral fins on either side of the fish's body (Fig. 4.8). Simple rowing movements of the pectoral fins represent a purely drag-based mechanism of propulsion (Blake 1981). Like ciliary or flagellar propulsion at low Reynolds numbers (see section 4.5), pectoral fin rowing requires a change in the projected area of the fin relative to the direction of fluid flow in order to alter the drag produced during the propulsive and recovery strokes. During the propulsive stroke, the fin is retracted back with the fin plane oriented *perpendicular* to its movement (Fig. 4.8(a), right), maximizing drag and hence the reaction thrust force that the water exerts on the fin. During the recovery stroke, the fin is rotated ('feathered') *parallel* to the flow (Fig. 4.8(a), left) and, in some fish, reduced in breadth, so that its projected area is minimized when the fin is protracted forward, keeping drag low. This enables the fish to achieve a net propulsive force in the forward direction.

Recent work using particle image velocimetry (PIV) to visualize and quantify flow has provided new insight into the hydrodynamic function of pectoral and caudal fin propulsion in fish. PIV uses a laser-generated 'light sheet' to illuminate a thin plane of water. By seeding the water with neutrally buoyant particles, detailed patterns of flow can be tracked and local flow velocities

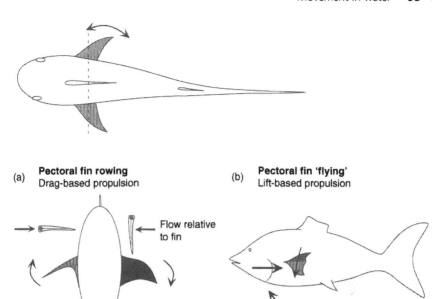

(a) **Pectoral fin rowing**
Drag-based propulsion

(b) **Pectoral fin 'flying'**
Lift-based propulsion

Flow relative
to fin

Recovery
stroke

Propulsive
stroke

Posterior view

Fig. 4.8 Many fish use their pectoral fins for swimming. Pectoral fin propulsion can be either (a) drag based, in which the fin is used for rowing, or (b) lift based, in which the fin is used much like a wing or the tail fin of a tuna. In many cases, both drag- and lift-based propulsion may occur over the course of the fin's movement.

quantified using high-speed videos of the fluid movement (Wilga and Lauder 2000). This computationally intensive experimental approach uses the same principles of tracking dye to map streamlines, but allows fluid forces to be calculated directly from the detailed pattern of flow velocities obtained from individual particles. This also enables investigators to examine non-steady, as well as steady, flows associated with particular locomotor mechanisms. Using this type of analysis, Drucker and Lauder (2000) have recently quantified the thrust produced by the pectoral fins of a swimming sunfish (Fig. 4.9), showing that donut-shaped rings of vorticity are shed at the end of each propulsive stroke. Their analysis indicates that in addition to a large posterior component of thrust, sunfish also generate a large lateral hydrodynamic force which is probably important to their stability and maneuvering ability.

It also seems likely that the unsteady effect of acceleration reaction due to the varying velocity of the fin as it is accelerated and decelerated during the stroke (Daniel 1984) may be an important feature of pectoral fin propulsion. Therefore the magnitude of thrust that is produced will depend on the rate of change of fin velocity, fin shape and stroke angle. Certainly these affect

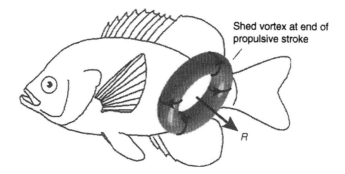

Shed vortex at end of
propulsive stroke

R

Fig. 4.9 The pectoral fin of the sunfish produces a shed vortex ring (similar to the shed vortices produced by the wings of birds and bats during slow flight; see Chapter 5). The rotational momentum of fluid (curved arrows) in the ring produces a net resultant propulsive force R. Quantification of fluid movement was achieved by means of PIV. (Reproduced from Drucker and Lauder (2000), with permission from the Company of Biologists Ltd.)

the magnitude and orientation of thrust that the pectoral (as well as tail) fin produces. The relative importance of these, and other, unsteady effects (see below) as propulsive mechanisms currently remains uncertain. However, it seems likely that they will prove to play a role, given the unsteady nature of fin movement.

In addition to drag-based thrust, paired pectoral fins can also generate thrust by means of lift produced when the pectoral fins are used as hydrofoils rather than as simple paddles (Fig. 4.8(b)). In this case, the fins are moved primarily in a dorsoventral plane and used as 'wings' (this is in contrast with sharks, which hold their pectoral fins fairly steady so that lift is generated mainly as a result of the shark's own forward motion (Fig. 4.7(c))). Lift is generated as a force acting perpendicular to the direction of flow over the laterally projected fin. By moving the fin downward with respect to the oncoming flow of water due to the fish's forward velocity, a lift force is generated that has a forward component acting to overcome drag. Therefore, just as for caudal fin propulsion in the tuna, lift provides net forward thrust. By rotating the fin in the opposite direction during the upstroke, the horizontal component of lift is maintained in the forward direction of the fish's travel (the vertical components of lift and drag cancel out during reciprocal downward and upward motion of the fin). Consequently, lift-based propulsion has the advantage of providing thrust over the entire pectoral fin cycle (as it does in caudal fin and fluke propulsion). In contrast, drag-based propulsion can only provide thrust over 50 per cent of the cycle. Generally, drag-based propulsion is favored at slow speeds of swimming and lift-based propulsion at faster speeds (owing to the increased flow achieved by the animal's forward movement in combination with fin's own movement). Nevertheless, the simple dichotomy of drag- versus lift-based propulsive mechanisms for slow versus fast swimming in fish that rely on paired pectoral fin propulsion can be misleading, as both mechanisms are likely employed during different phases of the fin stroke. Consequently, changes in fin kinematics and the relative importance of these two mechanisms are likely to occur over a range of swimming speed within a species, as well as when comparing pectoral fin morphology and locomotor function among different types of fish.

In addition to caudal and pectoral fin propulsion, many fish also use elongated dorsal and anal fins for swimming. These fins undergo lateral undulations that are propelled as a traveling wave down the length of the fin. Consequently, they are believed to provide propulsion similar to the drag-based mechanisms associated with whole-body undulation in anguilliform fish and eels.

4.5 Jet-based fluid propulsion

Jet propulsion has evolved a number of times in a diverse array of animals, resulting in a rather eclectic assortment that includes many types of invertebrates: tunicates and ctenophores (Vogel 1996), jellyfish (DeMont and Gosline 1988), scallops (Marsh *et al.* 1992) and, most notably, cephalopod molluscs (*Nautilus*, cuttlefish and squid). Squid are best known for their prowess as jetters, reaching speeds of up to 8 m/s, and have been described as 'invertebrate Olympians' because of their explosive mode of locomotion (O'Dor and Webber 1991). The general design of jet-propelled organisms involves a mechanism for ejecting a bolus of fluid at high velocity from an internal body cavity. Most jetters do this by means of a muscle-lined wall which encloses a fluid-filled chamber. Once again, thrust is achieved in reaction to the momentum of the fluid discharged by the jet. Contraction of circumferential muscle fibers decrease the diameter of the chamber, causing the fluid to be ejected through a port. The direction of the port allows the animal to orient its jet and thus control the direction of its movement. In the case of scallops, jetting is achieved by the contraction of an adductor muscle which causes the two valves (shells) of the animal to close against the mantle cavity. In most animals, an antagonistic set of muscles (longitudinal or transverse) cause the cavity to expand and refill for the next jet. In the case of scallops, however, an elastic hinge pad (abductin) serves as a compression spring that re-opens the valves.

Despite the fairly simple design of jet-propelled animals, jetting is not a common mode of locomotion. This is because jet propulsion is inefficient when the fluid must be contained within and expelled from a body cavity. The basic problem lies in the ability to generate thrust by means of momentum discharge (power output) versus the kinetic energy of the discharge (power input). As for swimming fish, the thrust T of a jetter integrated over a complete jetting cycle (from eqn (4.6)) is given by

$$\int T \, dt = MV \qquad (4.7)$$

where M and V are the mass and velocity of the water ejected in each jet. Whereas momentum thrust depends on V, the kinetic energy of the discharge depends on V^2. Consequently, jet propulsion is most efficient when the mass of fluid accelerated rather than the velocity of the jet is increased. Greatest efficiency is achieved when the jet velocity approaches the animal's forward velocity. This can only be achieved by very-high-frequency low-amplitude jets. This is the basis of jet-propelled engines in aircraft, but is not a solution for biological organisms. Instead, they must contain the mass of fluid within their bodies, which severely limits their efficiency. Although jetting represents an effective means for rapid acceleration (and hence predator avoidance or predatory strike), it is a costly mode of travel. O'Dor and Webber (1991) calculate that it costs squid about twice as much to go half as fast as a fish of the same size.

4.6 Movement at low Reynolds number: the reversibility of flow

Life at low Re is a particularly sticky business compared with life at high Re. Because viscous forces dominate at low Re, inertia can be ignored. At low Re, boundary layers are relatively thick because velocity gradients are slight, turbulence is absent and, since mixing across streamlines does not occur, flows are fully reversible. The symmetrical oscillatory movements of fins and body surfaces that larger aquatic organisms use to propel themselves through the water will not work under these conditions. If the cilia of a paramecium or the flagella of a bacterium were to move back and forth symmetrically, the animal would simply move forward and then back to its original position during each propulsive and recovery stroke. No net forward progression would be achieved.

Consequently, at low Re there must be some asymmetry in the kinematic *shape* of the propulsive and recovery strokes for any net thrust to be produced. Moving the cilia fast during the propulsive stroke and slow during the recovery stroke will not work either. This would only affect the rates of forward and backward movement. The end result would be the same—the organism would be in the same place at the end of a full cycle of movement. Because viscous shearing of the fluid is the sole mechanism available for generating thrust, low Re animals achieve net forward thrust either by changing the shape of their propulsors to maximize drag for thrust production during the power stroke and reduce drag during the recovery stroke (ciliates), or by taking advantage of helical or propagated bending waves (flagellates). Because drag depends solely on the viscous interaction of the organism with the surrounding fluid (i.e. pressure drag is unimportant), the overall shape of the animal's body is largely irrelevant to its ability to move through the fluid. Consequently, the shapes of these organisms are quite varied and no evidence of streamlining to reduce pressure drag is observed. Indeed, streamlining is undesirable because it tends to increase skin friction drag by increasing the surface area of the body exposed to flow.

4.6.1 Flagellar swimming

With the exception of bacterial (prokaryote) flagella which function as biological rotors, all eukaryote flagella and cilia represent cellular projections that consist of a $9+2$ arrangement of microtubules linked together by various microtubular-associated proteins (Fig. 4.10(a)). Bending of the flagella and cilia

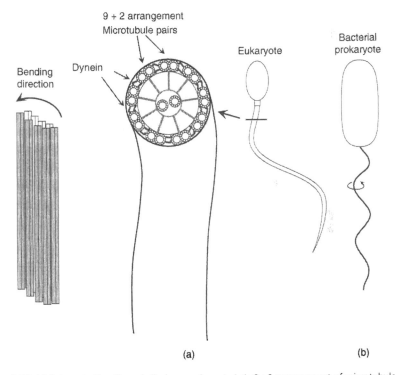

Fig. 4.10 (a) Eukaryote flagella and cilia have a characteristic $9+2$ arrangement of microtubules. Bending of the microtubules and hydrodynamic force are produced by the motor protein, dynein. Many dynein molecules bind at different sites between adjacent microtubules along their length causing them to slide with respect to each other. This induces an overall bending of the flagellum or cilium. (b) Bacterial flagella also are comprised of microtubules but have a different arrangement. In addition, prokaryotic flagella rotate at their base rather than passing waves of bending along their length.

occurs by sliding of the microtubules relative to each other, much like the relative sliding of actin and myosin filaments in skeletal muscle. This is driven by the molecular motor **dynein**, which is activated to produce the force needed to translate the microtubules with respect to each other. The circumferentially paired microtubules overlap so that one is incomplete. This asymmetry ensures that the tail of the dyenin molecule is anchored to the A microfilament of the pair, allowing the head end of the dynein to interact with the B microfilament of the adjacent pair. ATP-dependent movement of the dynein toward the minus end of the B microfilament is similar to the movement of dynein along microfilaments within cells for the purposes of cell transport (see Chapter 6) . The local bending moments generated by microtubule translation induce planar or helical bending waves that are transmitted along the length of the flagellum. Ciliary bending tends to be mainly planar. Despite differences in wave kinematics, flagella and cilia are morphologically similar and of fairly uniform diameter (0.2 μm) across a wide range of taxa. However, they can vary considerably in length (10–1000 μm) and the number of waves (one to four) transmitted.

Prokaryotic flagella (Fig. 4.10(b)) are morphologically and kinematically distinct from eukaryotic flagella and cilia. They are typically much smaller (0.02 μm in diameter and up to 20 μm long), less flexible and lack the characteristic internal 9 + 2 microtubular arrangement. The flagellum itself is a helical tube consisting of a single protein subunit **flagellin**. Prokaryotic flagella function by rotating at their base rather than by bending along their length. Rotation is driven by a proton gradient established across the plasma membrane of the flagellum organelle and the inside of the bacterium (the details of which are beyond the scope of this book). In fact, bacterial flagella represent the only true biological rotary devices known. Helical rotation of the rest of the flagellum is believed to result from passive transmission of the forces produced at the base. Bacteria, such as *Escherichia coli*, swim toward chemical attractants and away from repellents. Because the flagella have intrinsic 'handedness', they draw together as a coherent bundle when the flagella rotate in a counterclockwise direction but splay apart when the flagella rotate in the opposite direction. This results in either effective swimming in a fairly steady direction or a 'tumbling' behavior that produces a more chaotic motion. In the absence of a chemical signal, flagellar rotation reverses every few seconds. This causes the bacterium to tumble and change direction randomly over time, so that interspersed 'bouts' of swimming and tumbling constitute a random walk. In the presence of a chemotatic stimulus, tumbling is suppressed and the bacterium swims more steadily toward the attractant signal using a 'biased random walk' (Berg 1983).

Regardless of the detailed kinematics of motion or the molecular mechanism of power generation, flagellar thrust is produced by the same mechanism.

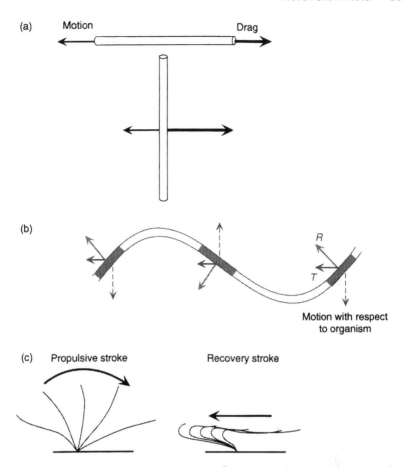

Fig. 4.11 (a) When a cylinder is perpendicular to the flow the drag that it encounters is 1.8 times greater than when it is oriented parallel to the flow. This difference in drag due to orientation enables (b) flagella and (c) cilia to generate a net propulsive force *T*. By orienting the cilia parallel to the flow during the recovery stroke much lower drag is produced than when they beat in a more perpendicular orientation during the propulsive stroke. Drag-based propulsion is the sole means of achieving net forward thrust at low Re.

A flagellum can be modeled as a cylinder. The drag experienced by a cylinder perpendicular to the flow (Fig. 4.11(a)) is about 1.8 times greater than the drag of a cylinder oriented parallel to the flow. By varying the angle at which segments of the flagellum move relative to the fluid (Fig. 4.11(b)) a net drag-based propulsive force *T* is generated perpendicular to the direction of the segment's motion (similar to that for undulating fish and eels (Fig. 4.4)). This is analogous to the motion of a cylindrical object that is pulled through the water at an inclined angle, which induces a slewing force (lateral to the direction of pull) that is equivalent to the thrust force generated by a flagellar segment.

The thrust developed by individual segments of the flagellum along its entire length is summed to yield the overall thrust generated. For helically translating and rotating flagella, thrust can be generated along most of the length of the flagella, but the animal's body will tend to corkscrew in the opposite direction. Planar undulation of flagella restricts effective thrust to those regions of the flagella that are inclined (>20°) to the direction of flow, but does not tend to destabilize the motion of the rest of the organism.

This analysis of flagellar propulsion at low Re ignores interactive effects of fluid movement past the body of the organism and its flagellum, as well as between adjacent ciliary propulsors. This latter limitation is a problem when considering closely packed cilia near the surface of an organism. Because of the much greater size and length of the flagella compared with the bodies of most flagellates, the drag on the head or body of the organism is generally less than 10 per cent of the flagellum and can be reasonably ignored. For bacteria, the requirement that rotation of the flagella be conserved by angular rotation of the head suggests that rotational propulsion is only effective if the radius of the head (modeled as a sphere) is greater than five times the radius of the flagellum.

4.6.2   Ciliary swimming

Ciliary propulsion involves the coordinated beating (metachrony) of many hundreds to thousands of short cilia (15 μm) along the surface of the organism. Ciliates generally have much larger body sizes (25–1000 μm) than flagellates and are able to achieve much faster speeds (where sizes overlap, ciliates move about 10 times faster (Sleigh and Blake 1977)). In contrast with the sinusoidal or helical motion of flagella, cilia beat in an asymmetrical fashion (Fig. 4.11(c)). During the power stroke the cilia are stiff and extended, oriented much like a cylinder perpendicular to the flow. During the recovery stroke they are flexible and bent near their base. This allows them to slide parallel to the direction of flow, reducing their drag by as much as 50 per cent. The twofold difference in drag during the power and recovery strokes accounts for much of the net thrust production in these animals.

However, cilia do not act as independent propulsors when closely packed. Instead hydrodynamic interactions among cilia probably occur. More complex hydrodynamic models have been developed in an attempt to characterize the propulsive efficiency of closely packed cilia (reviewed by Daniel *et al.* (1992)). For closely packed cilia, the summed interaction of the cilia can be reasonably predicted by treating the ciliary tips as a continuous surface of the organism and analyzing the overall metachronal wave of ciliary motion that this surface generates to produce hydrodynamic thrust. For less closely packed cilia, other models have been developed to take account of flow between cilia in order to calculate a more detailed velocity profile of flow within the ciliary field, but these are also beyond the scope of this book.

How effective are flagella and cilia? One measure of the effectiveness of fluid propulsion is **hydrodynamic efficiency**, which measures the rate of useful work performed relative to the total rate of work done by the propulsors. Useful work represents a measure of the distance that the animal moves for a given propulsive force. Hydrodynamic efficiencies of 0.09–0.28 have been estimated for flagellar propulsion (Daniel *et al.* 1992), whereas a value of 0.25 has been estimated for ciliary propulsion (Vogel 1994).

4.6.3   Size considerations

Why are ciliates generally so much larger than flagellates? What sets the limit to the size of cilia and the bodies of ciliates? Scaling arguments (Sleigh and Blake 1977) suggest that the rates of working and bending moments (proportional to l^3) increase faster than the cilium's resistance to bending (proportional to l^2) and much faster than the fluid velocity generated at its tip (proportional to l). Consequently, there is a limit to how long a cilium can be without compromising its structural and energetic effectiveness as a propulsor. Such limits do not apply to flagella because they do not bend as a whole but pass waves of bending along their length. The solution for larger body size is to have many short cilia or one or two long flagella. The disadvantage of flagella is that speed seems to be compromised, which may explain why flagellates are generally much smaller than ciliates. Vogel (1996) also suggests that cilia may be advantageous for facilitating diffusive exchange by moving stagnant fluid away from the body surface and for being less disruptive to the surrounding fluid environment which would give less signal to either predator or prey.

Limitations to larger body size seem to be set, at least in part, by hydrodynamic efficiency. Hydrodynamic efficiency is predicted to decrease with increasing size for both ciliates and flagellates. By assuming a constant ciliary density over the body surface and uniform propulsive efficiency of individual cilia, the overall hydrodynamic efficiency of ciliates is predicted to scale inversely proportional to the animal's body length (proportional to l^{-1}). Consequently, an upper limit to body size by means of ciliary propulsion appears to be around 0.1 mm.

4.7   Air–water interface: surface swimming, striding and sailing

Surface tension and waves become important factors influencing the locomotion of animals at the water surface. Fish typically swim underwater rather than at the surface (except for those species which leave the water to fly or to strike their prey). When swimming, most aquatic mammals and some birds also swim underwater, but many swim at the surface and all must come to the surface to breathe. Animals which move at the surface encounter additional drag resulting from wave formation. The drag at the surface can be as much as five times greater than the drag experienced at a greater depth. As we discuss in Chapter 9, this increases the metabolic cost of swimming at the surface.

Wave formation incurs a drag penalty because it involves work to elevate a mass of water against gravity. The Froude number ($Fr = v^2/gl$), introduced in Chapter 3, section 3.8, was originally defined by a marine engineer (Froude) to express the relative importance of a ship's inertia versus its wave drag. At a low Fr (i.e. low velocity and long length) only small waves are produced. At a higher Fr wave-induced drag increases, reaching a maximum at $Fr = 0.45$. Above this value, drag decreases because the boat moves at a speed great enough to plane over the water surface. Planing is probably a rare event for animals, but is certainly used by water birds when they land from a flight and appears to be used by ducklings when they swim. For boats and animals that are unable to plane, 0.45 represents their limiting performance; more propulsive energy will only produce larger waves, not higher swimming speeds. Generally, low speeds are favored at the surface. At low speeds, the bow wave created in front of the ship moves out and away at a faster speed than the ship, so that the ship remains level. However, at faster speeds, the ship eventually moves at a greater speed than its bow wave, causing it to 'swim' uphill.

The world record for the 100 m freestyle, held by Matt Biondi, is 48.6 s or about 2 m/s. To put this in perspective, a person walking fast along the side of the pool could readily match this speed. Quite obviously our athletic endurance and performance on land outstrips our abilities in water. Interestingly, this is also about the sustained cruising speed of many large fish. For Biondi, this represents a swimming speed of approximately one body length per second, making his Fr about 0.45. Consequently, Biondi's time represents the top performance for someone of his height. In general, surface swimmers must move at rather low speeds in order to be economical, in the range of $Fr < 0.2$. The maximum speed of a mallard duck with a hull length of 0.3 m should be about 0.7 m/s. This is quite low compared with a fish of similar size, which could easily achieve speeds of 2 m/s or more when swimming at depth. Similarly, muskrats swimming on a pond rarely exceed speeds of 0.6 m/s ($Fr = 0.16$). Surface swimming by ducks and muskrats provides clear examples of drag-based propulsion. In addition to overcoming drag, it is certainly the case (as for fish that are pectoral fin rowers) that these animals must produce additional thrust to overcome the acceleration reaction of water that is propelled backward by their feet. The relative importance of drag versus acceleration reaction to thrust production remains to be determined.

4.7.1   Striding and sailing on the water surface

A few small invertebrates, notably water striders and fisher spiders (Fig. 4.12(a)), take advantage of their size by using the surface tension of water to support themselves. Surface tension (N/m) equals a force exerted per unit distance; it essentially represents the work (N × m) required to deform a liquid over a unit area (m^2). The surface tension of water decreases only slightly with increased temperature but increases with salinity (it is 36 per cent greater in seawater than in fresh water). Water striders use surface tension to generate forces equal to or greater than their own weight in order to step over the surface of ponds or streams. The force exerted upward on a leg is equal to the surface tension of the water multiplied by the wetted perimeter of the leg. This force acts tangent to the water surface. Consequently, as the leg sinks further into the water its line of action becomes more vertical, enhancing weight support (Fig. 4.12(b)). It has generally been thought that, in order to move forward, a rearward push of a leg causes an asymmetrical reaction force from the fluid surface, giving the animal a forward acceleration. However, a recent study of fisher spiders (Suter *et al.* 1997) found that when these animals row across the surface they use *drag resistance* of the moving leg in combination with the dimple that the leg creates on the water surface (Figs 4.12(a) and 4.12(c)) to exert a propulsive force. Hence surface tension and not forward movement is apparently most important for weight support. In a subsequent study, Suter and Wildman (1999) found that the ability to maintain the integrity of the dimple was size and speed dependent (varying with size but inversely with speed). As a result, faster moving and/or smaller spiders switch from rowing, in which four limbs are used to propel the animal along the water surface, to a galloping gait, in which six limbs are used to propel the animal into the air in successive strides (Fig.4.12(d)). By doing this, the spider is released from the constraint of having to maintain contact with the water and the integrity of leg–surface dimple interaction. The recovery stroke occurs while it is airborne, allowing it to prepare its limbs for the next propulsive support phase.

Whirligig beetles use an approximately 50:50 balance of surface tension and buoyancy to support themselves, with their appendages providing paddles for drag-based propulsion similar to fisher spiders. Finally, stoneflies use surface tension to glide across the surface of a pond, with their wings raised as sails to provide propulsion. Indeed, Marden and Kramer (1994) have argued, based on studies of stoneflies which use their wings to sail,

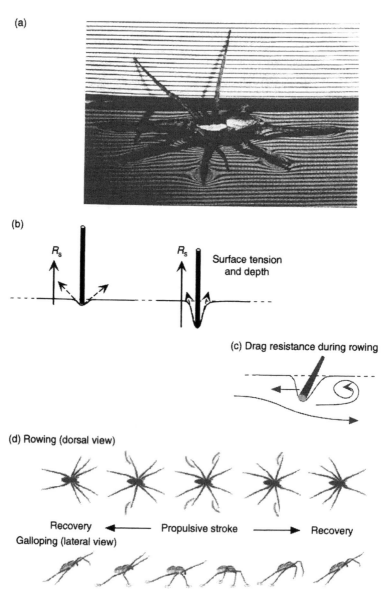

(a)

(b)

R_s R_s Surface tension and depth

(c) Drag resistance during rowing

(d) Rowing (dorsal view)

Recovery   ←   Propulsive stroke →   Recovery

Galloping (lateral view)

Fig. 4.12 (a) A fisher spider buoyed by surface tension on the water's surface. (b) Surface tension produces a net upward resultant force R_s which increases with the depth of the unwetted limb in the water. See text for further details. (c) Asymmetry in the surface tension acting on the limb can be used by the spider and other surface striders to produce drag-based rowing propulsion when moving at slow speeds. (d) To move faster, the fisher spider actually leaps into the air between periods of limb support. It also reduces the number of limbs that are in contact with the water surface, enabling reduced drag and faster movement. (Reproduced from Suter *et al.* (1997) and Suter and Wildman (1999), with permission from the Company of Biologists Ltd.)

that sailing provides an attractive scenario for how selection might have favored the evolution of a winged appendage in insects. Obviously, the first wing-like appendages would have been too small to provide adequate lift for flight. Sailing provides a plausible intermediate role for such an appendage.

The size range for effective use of surface tension is quite limited. If an animal is too large, the water cannot support its weight. If it is too small, the stickiness of the fluid, its viscosity, becomes a problem and the tiny beast cannot overcome the fluid's surface tension with its own inertia. The switch to a galloping gait by fisher spiders suggests a means of avoiding this problem at small size—by becoming airborne in between support phases of the stride. At the large end of the size range, a 70-kg man would need to have feet more than 30 m long in order to support his weight while walking on the surface of water!

4.7.2   Running on the water surface at large size: integrating terrestrial and aquatic lifestyles

Aside from a biblical accounting of Jesus Christ walking on the surface of water, there is at least one other biological example of a vertebrate that can manage the feat, albeit not at a walk. Iguanid basilisk lizards (known popularly as 'Jesus Christ lizards') have been observed to run across streams and ponds (for predator escape) in the Central American tropics with their head and trunk elevated above the water (Fig. 4.13). These animals are far too large to employ surface tension for support. Instead, they take advantage of the mass density of water, which exerts a reactive force when accelerated rapidly (the acceleration reaction force introduced in section 4.3.1). Basilisks achieve weight support by running rapidly with webbed feet which produce both an acceleration reaction force when their foot slaps the water surface and a cavitation force that acts upward, balancing their weight (Fig. 4.13(b)) (Glasheen and McMahon 1996). Just as for terrestrial running, the peak forces generated by this slapping mechanism must exceed the animal's weight to compensate for periods in which lower forces are exerted. By entraining a cavity of air, the foot generates a buoyant force (proportional to the entrained air volume) in addition to that generated by the acceleration reaction of the water below. The air cavity also allows the foot to be picked up out of the water with minimal resistance. Small basilisks are capable of literally running out of the water from a submerged position. Larger lizards are more cumbersome and perform best when they race from a bank, landing on the water at high speed. Observations of small green iguanid lizards suggest that they, too, may be able to run on water. But for both lizard species, body size plays a critical role in the use of this amazing locomotor behavior.

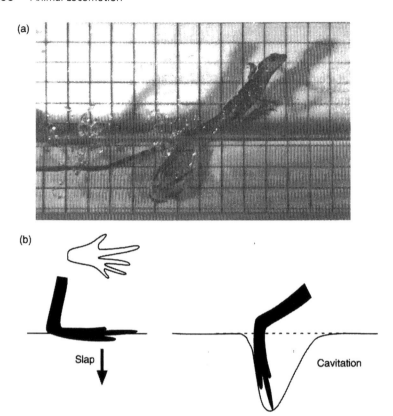

Fig. 4.13 (a) At a much larger size, a basilisk lizard runs over the water surface (b) by initially slapping the water surface to generate an impact reaction force and subsequently generating a cavitation reaction force produced by entraining air which displaces fluid from the foot cavity. This also allows the lizard to withdraw its foot with minimal drag from the air cavity produced.

4.8 Muscle function and force transmission in swimming

In contrast with terrestrial locomotion, in which some muscles can function to absorb and return energy to a running or bouncing animal by generating force under nearly isometric conditions (facilitated by elastic savings in tendons), the muscles which power swimming almost exclusively shorten during their contraction to do positive mechanical work. This is true of the muscles of both vertebrate and invertebrate swimmers. For most (i.e. non-jetting) swimmers, deceleration of a moving appendage can be achieved passively through drag, rather than requiring the active absorption of energy by muscles. In many fish and cetaceans the body musculature of the animal's trunk are the primary propulsors. In crustaceans, such as shrimp and lobster, it is the abdominal musculature that flips the animal's tail for propulsion. In other crustaceans, such as crabs, and amphibians, such as frogs, the limb muscles are used for swimming. Because most of the work and our understanding of muscle function during swimming has focused on fish, we shall concentrate here on fish axial muscle organization and function. However, the general principles that govern the function of fish musculature are also likely to apply to the swimming muscles of other animals.

In fish, the axial muscles (myomeres) undergo sinusoidal oscillations of length change (as evidenced by kinematic displacements of the body axis and, more recently, by means of direct recordings of muscle length change, e.g. Shadwick *et al.* 1999) generating fluid propulsive forces by means of the undulatory movements of the body surface, particularly those of the posterior half of the body and the caudal fin. The sinusoidal pattern of length oscillation relative to force development in fish axial body muscles has been studied extensively in recent years using the work-loop technique (Chapter 2). Using this approach, the effects of swimming speed (i.e. cycle frequency), length change, timing of activation, and temperature on the ability of muscles to generate mechanical power have been investigated. During swimming the axial musculature on the convex side of the body's bend is activated to generate force just before or as the fibers finish being stretched. This allows the fibers to develop force rapidly (Figs 2.5(a), 2.5(b) and 2.6), and subsequently causes them to shorten and the animal's body to bend in the opposite direction (becoming concave). This local force is either transmitted posteriorly down the length the animal's body to the tail or is transmitted directly to the fluid adjacent to the body surface for propulsion. Thus muscle work is transformed into hydrodynamic work. Traveling waves of bending and fluid propulsion are driven down the animal's body in this way by means of a transmitted wave of electrical activation by the nervous system to the muscles along the body axis.

The conical arrangement of myomeric muscles along the body axis of fish represents a functional morphologist's puzzle of force transmission. Forces developed by the muscles are anchored by or transmitted to the vertebral column and to the tail. In sharks, a cross-fibered array of collagen fibers beneath the skin is also used as an 'external body tendon' to which the muscles attach, allowing the shark to transmit force posteriorly with a large mechanical advantage to bend its body and ultimately its tail (Wainwright *et al.* 1978). In tuna, well-developed lateral tendons emanate from the axial muscles to attach to the tail. The presence of these large tendons is strong evidence that they transmit significant forces and corresponds to the emphasis on caudal fin propulsion by these fish. Recordings of forces from these tendons confirm their role in powering oscillation of the tail (Knower *et al.* 1999). At present, however, the various pathways and relative importance of force transmission by fish myotomal muscle remain unclear. The complex arrangement of these muscles and the many connective tissue and skeletal components to which they attach makes their study a fascinating but ongoing challenge.

In most fish, two distinct populations of fibers exist within the axial musculature: the red fibers are slow oxidative, whereas the white fibers are fast glycolytic. Typically, in most fish the red fibers constitute a relatively small portion of the myotome and are found in a longitudinal band near the body surface. In certain specialized 'regionally endothermic' fish (tuna and lamnid sharks) (Fig. 4.14) substantially more red muscle is found closer to the backbone, within the larger white muscle component. These fish are capable of maintaining their deeper red muscle at a warmer temperature than the surrounding water (Carey 1973). At slow swimming speeds only the red SO fibers are activated to produce a slow traveling wave of propulsion. As swimming speed increases, white muscle fibers are also recruited to power the animal's swimming, providing faster and stronger waves of propulsion. However, because the white (FG) fibers are less oxidative than the red (SO) fibers, endurance is progressively reduced at faster swimming speeds. The locomotor performance of most fish (which are ectotherms) is strongly affected by water temperature. Interestingly, at low temperatures, several fish begin to recruit faster contracting white fibers at slower swimming speeds in order to compensate for reduced muscle power output at these low temperatures (see Chapter 10, Fig. 10.11) (Rome *et al.* 1984). In regionally endothermic fish, such as tuna and billfish, counter-current heat exchangers in the vascular system supplying the muscles maintain internal red muscle temperatures as much as 10 °C higher than the surrounding water. By keeping their red musculature warm, the fish are able to achieve greater endurance and faster swimming speeds in colder waters.

4.8.1 Work and power output of red versus white muscle

By alternately shortening and lengthening to bend the fish's body and tail back and forth, the axial musculature of swimming fish does mechanical work during the shortening phase of its length cycle as it contracts to develop force. Studies of the neural activation of muscle segments at different sites along the length of swimming fish show that the muscles are typically activated just as they finish being passively lengthened, so that they develop force while shortening (see Fig. 2.6). This allows the muscles at any one location on the animal's body to do work (force × shortening distance). This work is transmitted posteriorly along the animal's body, to its tail, and is used to produce the hydrodynamic thrust necessary to propel the animal forward. Studies of the SO red muscle fibers from fish (Rome *et al.* 1993) indicate that they shorten at a velocity that is close to their optimum ($0.3–0.4V/V_{max}$) (see Chapter 2, Fig. 2.4) for doing work efficiently and generating power (work/time). Hence, fish typically cruise at a steady uniform speed using their red muscle to maximize their

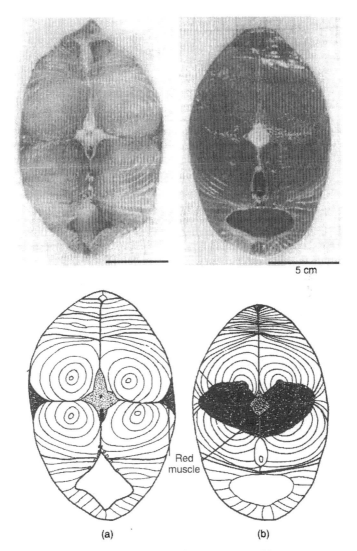

Fig. 4.14 Comparison of red and white axial muscle organization in (a) mackerel and (b) tuna (two scombrid fishes). In most fish, such as the mackerel, the red muscle represents a limited portion of the myotomal muscle and is located just beneath the skin lateral to the white muscle. In tuna and other fish which warm their red muscle, the red muscle is more extensive and lies deep to much of the white muscle. Counter-current heat exchange keeps the red muscle warmer than the water and the rest of the fish (some fish also maintain elevated brain and eye temperatures). (Reproduced from Westneat and Wainwright (2001), with permission from Academic Press).

swimming efficiency and reduce their cost of movement. When they accelerate or swim quickly they use their white muscle to maximize power output for escaping predation or to catch prey.

4.9 Summary

In this chapter we have seen that the physical properties of water, and the common hydrodynamic principles that emerge from them, govern the swimming performance and the diversity of aquatic propulsive mechanisms which animals have evolved. Although the buoyancy of water has enabled a tremendous range in the size of aquatic animals, body size still plays a crucial role in determining the physical regime of fluid propulsion which can be successfully employed. Whereas moderate and large animals must contend with drag resulting from their own inertia, the aquatic world of very small animals is governed by viscosity. As a result, the strategies which work for effective fluid propulsion at intermediate to large Re (streamlining and lift) do not work for very small animals. Instead, they must overcome the reversibility of flow at low Re by creating asymmetric patterns of drag-based propulsion, reminiscent of the strategies of both pectoral fin and surface rowers. Within a given Re regime, however, we find a spectacular diversity of body forms and evolutionarily successful hydrodynamic propulsors, ranging from lift-based caudal fins and the body axis of fish to the cilia and flagella of unicellular swimmers. Finally, as is almost always the case in biology, we also observe unusual, yet quite dramatic, forms of aquatic propulsion such as the jetting of squid, the surface running of basilisk lizards and galloping fisher spiders.

5 | Movement in air

The aerial performance of flying animals is remarkable and has inspired human myth and experimentation over much of our history. The grace and beauty of a heron in flight, the power and drama of a predatory attack by a diving hawk and the flitting maneuvers of a bumblebee or a hummingbird all capture the extraordinary performance of biological flying machines. For comparison with the man-made flying machines that they inspired, the flight of a house fly at 3 m/s represents a speed of 430 body lengths/s (Table 5.1). When normalized for size in this way, a fly achieves a speed that is more than 12 times greater than the speed of a high-performance fighter jet and 80 times greater than the speed of a propeller-driven airplane.

In addition to such spectacular performance, flight has proven a highly successful mode of life for a wide range of taxa, having contributed to the enormous success of insects (>800 000 species), birds (>8000 species) and bats (>850 species), which constitute the second most speciose group of mammals. Although more expensive than swimming, flight is a cheaper means of transport over a given distance than when moving on the ground (Chapter 8), particularly when changes in elevation must be negotiated. Flight enables animals to migrate and forage over large distances, avoid harsh environmental conditions (e.g. desert, ocean) and thereby reach otherwise inaccessible foraging sites. In addition, flight provides an exceptional means of predator defense as well as excellent access to prey and other food resources.

Aerial flight involves the same fluid-mechanical principles that underlie aquatic locomotion. However, because of the 800-fold lower density of air compared with water, important differences exist. Unlike swimming, support of body weight is the key problem when moving through the air. Consequently, the wings must produce lift to support the animal's body weight as well as produce thrust (Fig. 5.1). Because effective lift production in a low-density

Table 5.1

Aircraft	Mass (kg)	Length (m)	Speed (m/s)	Speed (lengths/s)
DH-60 'Moth' biplane	450	7.2	39	5.4
Cessna A-37B 'Dragonfly'	4500	8.6	219	25.5
Concorde	15 000	62.1	606	9.8
Airbus A340	240 000	63.7	253	4.0
MD F4 'Phantom' jet	20 300	19.4	639	32.9
Gossamer Albatross (human-powered plane)	100	10.4	8.0	0.8
House fly	0.0001	0.0007	3	430
Butterfly	0.0002	0.05	1	20
Starling	0.070	0.15	15	100
Duck	1.0	0.28	25	89
Swan	8.0	0.8	18	23

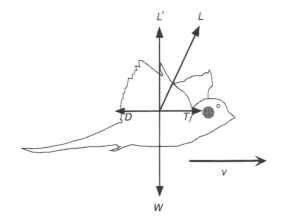

Fig. 5.1 Flying bird showing the balance of forces that act on it during powered flapping flight at a constant speed v (L, lift; L', component of lift to support weight, W; T, thrust; D, drag).

fluid, such as air, requires a high flow velocity, flying animals move at much higher speeds than swimming animals, or move their wings rapidly. As a result, the Reynolds number (Re) range for most biological fliers is sufficiently high (10^2–10^7) that inertial forces largely dominate. This means that pressure drag is a more important consideration than viscous (or friction) drag, except perhaps in the smallest fliers (e.g. fruit flies which weigh about 0.01 mg and operate in an Re range from 10 to 100).

5.1 Lift, drag and thrust in flight

As we have already noted for water movement past a hydrofoil, when an airfoil encounters an oncoming flow of air at an angle, the flow separates around the

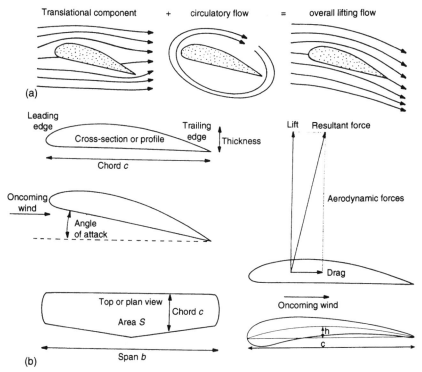

Fig. 5.2 (a) Asymmetry of airflow past an airfoil (being faster above and slower below) can be decomposed into a translational and a circulation component. Flow is described by streamlines. See text for details. (Reproduced from Vogel (1994), Figs 11.1 and 11.2, with permission from Princeton University Press.) (b) Airfoil shape and definitions of important shape and aerodynamic variables.

airfoil such that the air which passes over the top of it moves at a higher velocity than the air that passes below it (Fig. 5.2(a)). This results from the fact that the air must travel a greater distance per unit time due to asymmetrical flow developed relative to stagnation points ($v = 0$) at the leading and trailing edges of the airfoil. The angle that the airfoil presents to the oncoming flow is referred to as its 'angle of attack'. In addition, rather than being symmetric, airfoils usually are asymmetrical in shape with their convex upper surface being more curved than their lower surface (which may be nearly flat or slightly concave). **Camber** (h/c, Fig. 5.2b) is a measure of an airfoil's curvature. Combined with an asymmetrical shape, camber allows an airfoil to generate a greater velocity differential (and aerodynamic lift) for a given angle of attack compared with a non-cambered airfoil. Insects and hummingbirds have more symmetric and uncambered airfoils because they operate to generate lift during both the downstroke and upstroke. The fins of fish also tend to be symmetrical because they, too, generate useful hydrodynamic thrust during alternating tail beats.

The velocity differential produced by asymmetrical flow past the airfoil results in a pressure difference (low above and high below). According to Bernoulli's principle, this results from the faster moving air above the airfoil versus more slowly moving air below. This pressure difference induces a net upward force on the airfoil that acts perpendicular to the incident airflow and is termed **lift**. Lift can also be thought of as the net circulation generated around the airfoil resulting from the velocity differential (Fig. 5.2(a)). As for any fluid, air does not actually circle around the airfoil, but the asymmetric flow pattern can be considered to reflect the sum of a translational component and a circular component of airflow. The circulation developed along the length of the airfoil is shed at its tip as vortices (at which point real physical circulation of the air does occur). Shed vortices represent the momentum transferred to the air associated with lift generation. Increased airflow results in an increase in both translational and circular components. Consequently, faster air speeds create more circulation and hence greater lift. Lift actually varies in proportion to the square of incident air velocity ($\propto v^2$), which depends on the animal's air speed in combination with the velocity of its wing as it is flapped. The dependence of lift on the translational free-stream velocity and rotational velocity is formally defined by the Kutta–Joukowski equation

$$L = l\rho v\Gamma \tag{5.1}$$

where Γ is the magnitude of the circulation, l is the length of the airfoil over which the circulation develops, ρ is the density of air and v is the animal's velocity. (For the more ambitious and mathematically inclined reader, a readable but more formal discussion of circulation and aerodynamic lift is presented by Milne-Thomson (1966).) This means that an animal can generate more lift both by flying faster and by having longer wings. Similar to drag, lift can also be (and is conventionally) defined as

$$L = C_l\rho Sv^2/2 \tag{5.2}$$

where C_l is the lift coefficient (analogous to the drag coefficient). In this case, S represents the profile area of the wing (often referred to as its 'planform') (Fig. 5.2(b)). Like the drag coefficient, the lift coefficient depends only on shape, orientation and Re. However, the two coefficients depend in differing ways on these factors, which underlies much of airfoil design.

We now see that the resultant aerodynamic force acting on a wing can be distinguished as two basic components: lift (which always acts *perpendicular* to the incident direction of airflow) and drag (which always acts *parallel* to the airflow). For a given shape and Re, changing a wing's angle of attack (orientation) alters the amount of lift relative to drag that the wing experiences (see Fig. 5.4c below). Increasing the angle of attack initially increases the amount of lift relative to drag. However, beyond a certain angle of attack, lift begins to decrease as drag continues to increase. At a critically large angle of attack an airfoil will 'stall' due to flow separation along its upper surface, which causes a sharp (and sometimes catastrophic) reduction in circulation and drop in lift. However, under controlled circumstances, such as when a bird lands, an increased angle of attack leading to a stall is critical for enabling the bird to slow down (due to increased drag) and descend lightly. In fact, birds have evolved specialized devices (the **alulae**, small feathers which attach to the 'thumb' of the leading edge of the wing) which probably help to maintain a stable attached flow of air above the wing at angles of attack that might otherwise produce stall. Consequently, biological airfoils can have a much broader plateau than simple airfoils, which delays their stall (Fig. 5.4(c)). Two other interesting points emerge from Fig. 5.4c. First, positive lift can be generated by airfoils even at negative angles of attack if an airfoil has an asymmetrical shape (camber), which favors a faster flow velocity along the upper surface (a symmetrical airfoil operating with a negative angle of attack will induce a reversed circulation and hence experience 'negative' lift). The second is that a tangent to the curve drawn from the origin defines the angle of attack at which L/D is maximum. This represents the optimal performance that an airfoil can achieve. Maximum lift-to-drag ratios in the range of 10–18 have been reported for soaring birds (see section 5.3) such as falcons, condors and albatrosses (the latter having the highest L/D ratio due to their extremely long narrow wings). Lower L/D ratios, in the range 2–8, have been observed for the flapping flight of smaller birds and insects.

It is clear that lift acts in a direction that is favorable to counteracting a flying animal's weight, but how does lift generate thrust in order to overcome drag? For this to happen, lift must have a forward component. This is achieved by moving the airfoil at an angle to the direction of the animal's forward travel (see Fig. 5.6 below) so that the lift vector has both an upward and a forward component. This is the basis of **flapping** flight, in which the motion of the wing downward relative to the forward (horizontal) movement of the animal induces a net airflow around the wing that is elevated with respect to the horizontal. As a result, aerodynamic lift has a horizontal component (thrust) which overcomes the drag acting on the animal and a vertical component (L'; Fig. 5.1) that counteracts its weight. It is important to remember that lift always acts perpendicular to the resultant path of incident airflow past the wing, which results from the wing's velocity relative to the animal, as well as the animal's forward flight speed relative to any prevailing wind.

5.1.1 Aspect ratio

A key parameter which affects the lift-to-drag performance of an airfoil is its aspect ratio (AR) which is defined most simply by the ratio of tip-to-tip length (span b) of the two airfoils to their average width, or chord, c (Fig. 5.2(b)).

Because wings taper toward their tips, the mean wing chord is often difficult to define. Consequently, AR is often defined as b^2/S (the square of span divided by the profile area of the wings). Long narrow wings have high AR values (e.g. albatross, AR = 15), whereas short stubby wings have low ARs (e.g. sparrow, AR = 5.5). Generally, insects have low-AR wings compared with birds. High-AR wings enhance lift relative to drag and therefore are a common feature of birds that glide and soar. The chief advantage of a low-AR wing is improved maneuverability. Because size also affects maneuverability, small birds and insects with short stubby wings are far more maneuverable than larger gliding birds. Shorter low-AR wings are also beneficial to seabirds which 'fly' underwater to catch fish (e.g. diving petrel).

5.1.2 Wing loading

The ability to generate lift depends on the surface area of the wings (eqn (5.2)). Consequently, in addition to changing a wing's angle of attack, increased lift can be achieved by increasing wing area. For birds and bats, changing wing area is an important control device for adjusting lift during landing and maneuvering. Changes in wing area also occur during each phase of a wing beat cycle. The ability to collapse the wing during the upstroke, for example, is important for reducing drag, especially during fast flight. Aircraft are similarly designed with the ability, albeit to a much lesser degree, to alter wing area (as well as camber) during take-off and landing. The weight of a flier relative to the area of its wings (W/S) defines its wing loading. Wing loading provides a quantitative comparison of how much lift a unit area of wing must produce to support the animal's own weight and any cargo that it is carrying.

Differences in wing loading have important implications for flight performance. Slow-flying birds generally have large wings (low wing loading), whereas fast fliers have higher wing loading. In general, bats (Table 5.2) operate with lower wing loading than birds of similar size. This enables them to be extremely maneuverable for catching insects or negotiating dense foliage in search of fruit sources. In the other direction, the relatively small wings but high wing loading of ducks and geese requires that they fly fast in order to generate sufficient lift to support their weight. These birds also operate their wings with a small angle of attack, which helps to reduce drag at fast flight speeds (see below).

Wing loading introduces a basic problem of scaling. The need to produce lift can be expected to vary with an animal's weight, but the ability to generate lift at a particular speed depends on wing area. For geometrically similar fliers this suggests a scaling of wing loading as $M^{1/3}$. This means that larger fliers should have greater difficulty generating enough lift to support their weight, especially at slower flight speeds and during take-off. Clearly a size limit to

Table 5.2

Species	Mass (kg)	Wing loading (N/m^2)	AR
Vertebrates			
Wandering albatross	8.7	140	15
Herring gull	0.54	51	9.5
Diving petrel	0.14	64	7
Andean condor	10.0	101	7.5
Buzzard	1.0	33	5.8
Sparrow hawk	0.2	28	6.5
Mute swan	8.0	230	9.2
Canada goose	1.8	155	10.1
Mallard duck	1.0	113	9.1
Black grouse	1.0	85	5.9
Magpie	0.22	35	5.7
Starling	0.075	37	7.2
Budgerigar	0.035	34	7.2
House sparrow	0.028	26	5.5
Swallow	0.024	16	8.0
Hummingbird	0.005	32	8.1
Archeopteryx[a]	0.27	55	6.3
Pterosaur[a]	15	32	10.5
Rousettus bat	0.14	25	5.9
Fruit bat	0.014	12.3	6.5
Greater horseshoe bat	0.023	12.2	6.1
Little brown bat	0.007	7.5	6
Insects			
Fly	0.200	15.4	12.3
Bumblebee	0.0002	15.7	10.0
Butterfly	0.0001	0.9	2.6
Sphinx moth	0.0005	1.3	6.4
Dragonfly	0.0001	2.0	5.1

[a]Extinct species.

animal flight, using skeletal muscle as a motor, must exist. The largest living flying animals are kori and great bustards, weighing in at 13–16 kg. Although past extinct fliers (including birds and pterosaurs) may have evolved greater weights and sizes than the bustards, it is unlikely that a vertebrate capable of powered (as opposed to gliding) flight has ever existed that exceeded 25 kg in weight. One exception was the human-powered Gossamer Albatross, which crossed the English channel (36 km) in 1979, having been successfully engineered to achieve sufficient aerodynamic lift to support its human pilot/motor (total gross weight, 100 kg). With a wing area of 45 m², it had a wing loading of only 22 N/m², well within the range observed for various vertebrate fliers (Table 5.2).

It is also not surprising that larger fliers tend to have relatively larger wings than smaller ones. Consequently, wing loading does not, in fact, scale as strongly ($\propto M^{0.22}$) as predicted by geometric scaling. For comparison, a Boeing 747 jet transport has a wing loading of 6000 N/m², compared with 101 N/m² for an Andean condor, 26 N/m² for a house sparrow, 50 N/m² for a bumble-bee and 3.5 N/m² for a fruit fly (Table 5.2). Nevertheless, the scaling of wing area is insufficient to maintain a constant wing loading across different-sized species. Larger fliers compensate for their lower wing loading by generally flying at faster speeds. Because lift varies with the square of speed (eqn (5.2)), faster flight can readily make up for reduced wing loading. Gliding animals generally have lower wing loading than non-gliders.

5.2   Power requirements for steady flight

The aerodynamic power requirements for flight can be separated into three main components: induced drag, profile drag and body (parasite) drag. Induced drag and profile drag are drag components that operate on the wings themselves, whereas body drag is the pressure and skin friction drag associated with air that moves over the body surface and therefore represents the parasitic cost of having a body which must be carried by the wings. In order to move at a steady forward speed a flying animal must generate sufficient lift to support its weight and overcome drag (drag × speed = power). Because the animal's muscles must generate mechanical power to overcome each component of drag, each can be plotted as a separate power component as a function of flight speed (Fig. 5.3). Therefore the total aerodynamic power requirement for flight is the sum of these three components.

5.2.1   Profile and parasite drag

Profile drag results from pressure and skin friction drag operating on the wings. As expected from eqn (4.4) (Chapter 4), profile drag increases with the square of velocity. Profile drag also increases with increased angle of attack. Consequently, if the angle of attack is reduced at faster flight speeds (which larger fliers do to compensate for the adverse scaling of wing loading), the increase in profile drag does not increase quite as rapidly as shown. The parasite drag of the body increases in a similar fashion with flight speed. However, because the wings of most flying animals have greater surface area than the body, profile power is typically greater than the parasite power. This is especially the case for larger gliding birds (see below). Profile power is greater than

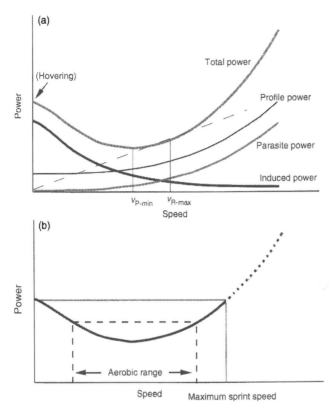

Fig. 5.3 (a) Aerodynamic power requirements versus flight speed. Induced power is highest during hovering (zero speed) and slow flight, declining as an animal flies faster. Profile and parasite power (due to drag acting on the wings and body respectively) increase at higher speeds. Parasite power increases most steeply because the velocity of the wings' motion, which determines profile power costs, increases more slowly with speed. These combine to give an overall U-shaped power curve, with a minimum at an intermediate speed ($v_{P\text{-min}}$). The tangent to the total power curve drawn from the origin defines the minimum cost of transport speed (or maximum range speed $v_{R\text{-max}}$).
(b) Diagram showing that the aerobic range of flight speeds for most birds is quite restricted. The maximum flight performance of most birds also limits their speeds to either very brief periods of hovering or short maximum speed sprints. Birds rarely perform at these limits and, as a result, may not operate with as much of a change in flight power requirement as suggested by a theoretical aerodynamic power curve.

parasite power at low speeds, because the wings must move at a much higher velocity than the bird's forward airspeed. As a result, the increase in profile power is less steep than the increase in parasite power, which increases with the cube of the bird's forward airspeed. In addition to the wings, some lift is probably also produced by the animal's body. Estimates for body lift range from about 5 to 20 per cent of the lift generated by the wings for a range of fliers which included bumblebees, locusts and zebra finches. Because of this, the width of the body is often included in the measurement of wing span.

5.2.2 Induced drag: the cost of finite wings

The reason that low-AR wings achieve lower lift-to-drag performance than high-AR wings results from the fact that the circulation which develops around the wing to produce lift is ultimately dissipated at the wing tip as a shed vortex. Longer wings maintain a greater proportion of the circulation bound to the wing compared with that which is lost as momentum when being shed from the tip as vortices. On balance, a longer wing (if sufficiently narrow) produces more lift than the additional drag incurred by increased length. In contrast with real biological and aircraft wings of finite length, infinitely long wings *theoretically* lose no energy due to tip vortices; momentum is lost only due to drag resulting from airflow over the chord-wise section of the wing.

The extra drag, which results in energy lost by finite wings at their tips, is referred to as **induced drag**. The air shed from the airfoil is referred to as the 'downwash', which has a downward component of kinetic energy. Therefore induced drag is also considered to represent the component of drag associated with lift generation which produces this downwash. This means that the product of induced drag and free-stream velocity (airspeed of the animal) equals the induced power cost for an animal to stay aloft with a wing of less than infinite span (or AR). Because a wing comes into contact with more air per unit time at faster flight speeds, but the lift required to stay aloft remains constant, less induced power is required at faster speeds (i.e. the downwash represents a smaller component of the airflow past the airfoil at faster speeds). Consequently, in contrast with profile and parasite power, induced power is high for hovering and slow speed flight but decreases inversely with increasing flight speed (Fig. 5.3). The increasing induced power requirement at progressively slower speeds explains why hovering flight is so difficult and energy demanding. When an animal hovers, all the circulation for lift must be generated by the flapping motion of the wings themselves. In contrast, in forward flight at faster speeds, the animal's own speed can be used to produce circulation, reducing the amount that the wings must generate via flapping. As flight speed increases, the induced power requirement for generating circulation continues to decrease.

Because of the decrease in induced power with increasing flight speed, which opposes the increases in profile and parasite power, the total power requirement for forward flight is considered to have a characteristic U shape (Fig. 5.3). In other words, total power is high at hovering and low speeds because induced power is high. As speed increases from low to moderate speeds, induced power declines more rapidly than the increase in profile and parasite power. As a result, total power decreases to a minimum, before increasing at faster flight speeds due to the rapid rise in profile power and, to a lesser extent, parasite power. The U-shaped power curve for flight has two interesting implications. First, it indicates that there is a particular speed at which it is cheapest to fly (minimum power speed $v_{P\text{-min}}$). Second, it suggests that that there is a speed at which the animal should fly if it wishes to cover the greatest distance as cheaply as possible (minimum cost of transport, or maximum range speed $v_{R\text{-max}}$). This speed is defined by the tangent to the curve drawn through the origin (which gives the minimum slope of power versus speed) and occurs at a higher speed than the minimum power speed.

It is important to note that the U-shaped power curve depicted in Fig. 5.3 is based on steady aerodynamic theory for 'fixed-wing' aircraft, in which the shape of the wing remains constant and airflow over the wing does not change through time. While these assumptions are reasonable for gliding and soaring flight, both are unrealistic for flapping flight. Consequently, changes in flight behavior and wing shape may be expected to modify the U-shaped power curve for flapping flight in different species. We shall return to this matter in section 5.5.2.

5.3 Gliding flight

The simplest form of flight to consider is gliding. This is because steady conditions of flow operate and aerodynamic theory can readily be applied. Gliding represents unpowered flight. Although the animal uses metabolic energy to maintain its wings extended, it generates no mechanical power with its flight muscles. Instead, gliders convert their potential energy into aerodynamic work, allowing them to cover a certain horizontal distance as they descend. During gliding the resultant of lift and drag forces acting on a wing exactly balances the weight of the animal, so that the animal descends along a fixed course at a constant speed (Fig. 5.4(a)). Under these conditions, the ratio of lift to drag (or C_l/C_d) equals cot θ (or $\tan^{-1} \theta$), where θ is the **glide angle**. Not surprisingly, birds which spend a great deal of their time gliding (see 'soaring' below) typically have high L/D ratios, achieved by having high-AR wings (see Table 5.1), which allow them to glide at small angles. Albatrosses, with AR = 15, have an L/D ratio of 20, allowing them to glide at an angle of 3° or less. Hawks and vultures have L/D ratios ranging from 10 to 15 (with a glide angle of 4° to 6°). By minimizing its glide angle, an animal maximizes its gliding distance. An albatross gliding from a height of 1 km above the ocean can travel 20 km in still air before reaching the water surface. Human-engineered sailplanes achieve an L/D of 40 (AR = 20), which enables them to travel 40 km for each kilometer of descent.

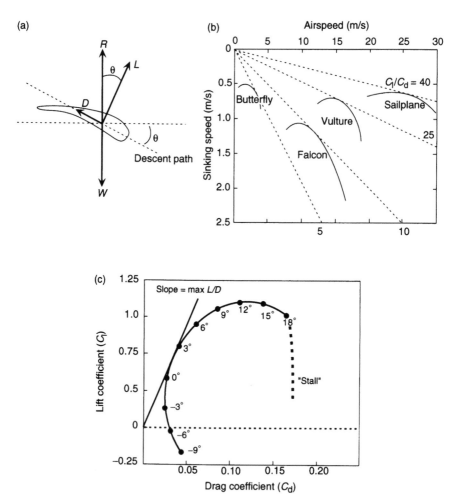

Fig. 5.4 (a) Lift–drag and weight balance during gliding flight and glide angle. (b) Glide polars (lift coefficient versus drag coefficient) for a sailplane, a butterfly and two bird wings. (c) The coefficient of lift (C_l) plotted versus the coefficient of drag (C_d) at various angles of attack for an airplane wing. The tangent to the curve from the origin gives the maximum L/D ratio for the airfoil and the angle of attack at which this is achieved. Similar L/D polar diagrams have been observed for bird and insect wings, but these are based on measurements obtained from a stationary wing. Consequently, reliable L/D graphs for animal wings as they move in flight are still unavailable. ((a) and (b) reproduced from Vogel (1994), Fig. 11.4, with permission from Princeton University Press.)

For arboreal gliders, such as a flying squirrel (which actually glides) or a gliding lizard, a smaller glide angle means that less vertical elevation is lost when gliding between trees. Nevertheless, these gliders have much lower L/D ratios (2.0 or less) than those of birds and bats. Gliding has evolved as an effective means of transport in a diverse array of arboreal animals including lemurs, opossums, frogs and snakes, in addition to the birds mentioned above. A distinction is commonly made between gliding ($L/D > 1$) and parachuting ($L/D < 1$), when the glide angle exceeds 45°. Animals which parachute typically exhibit less aerodynamic specialization.

By substituting $L = mg$ in eqn (5.2) the glide speed can be calculated as

$$v_g = (2mg/\rho S C_l)^{1/2} \tag{5.3}$$

which shows that with a lower wing loading mg/S or a higher lift coefficient an animal can glide at a slower speed. An animal cannot glide more slowly than the speed at which it would stall (maximum C_l). The slotted primary wing tip feathers of hawks and vultures represent one means of enabling these birds to delay stall and glide at slower speeds. In addition to horizontal range, the duration of a glide may also be important to an animal. Glide duration depends on the sinking speed of the glider, which is

$$v_s = v_g \sin \theta. \tag{5.4}$$

The plot of the sinking speed of a glider against its horizontal air speed $v_h = v_g \cos \theta$ (Fig. 5.4(b)) represents its 'glide polar'. The glide polar shows how a glider can alter its air speed versus its sinking speed by changing its angle of attack, wing camber and wing span. A tangent drawn from the origin (broken lines) gives v_s and v_h at the minimum glide angle. Changes in wing span, which drastically affect wing area and AR, are the most effective mechanisms for changing v_h and v_s. Gliders that wish to remain aloft as long as possible (minimize v_s) operate at the upper left of their glide polar. In contrast, a raptor that wishes to descend as fast as possible to catch its prey operates at the lower right end of its glide polar. It does this by retracting its wings back, reducing their span and angle of attack, in order to maximize its sinking and glide speeds (in addition to reducing its profile drag).

Because speed affects lift and drag similarly at moderate to high Reynolds numbers, glide angle is largely independent of speed. Consequently, heavy and light gliders with the same L/D ratio descend along nearly the same path. However, because weight is balanced by lift, which varies approximately with v^2, heavier gliders necessarily travel at faster speeds than light ones. Fast glide speeds can be a problem, and so gliders tend to be light weight. As a final consideration, scaling once again enters the picture. Smaller gliders have lower L/D ratios because they tend to have proportionately greater profile drag due to viscous effects at low Re. Consequently, small size indicates a steeper glide angle. Because of this, insects with $L/D < 2$ are not generally very good gliders.

5.3.1 Soaring

Soaring is specialized form of gliding flight in which a bird takes advantage of energy available in natural air-movement patterns in order to remain aloft for considerable periods of time without having to flap its wings regularly. Soaring allows these birds to gain substantial energy savings (estimated to be as high as 67 per cent) for travel, surveillance of prey or actual feeding. Two general forms of soaring are distinguished: static soaring and dynamic soaring. Static soaring involves 'slope soaring' and 'thermal soaring'. In the case of slope soaring, a wind moving uphill over a slope, as would be the case over the side of a hill, a cliff face or even an ocean wave, provides the energy to keep the bird aloft. Glide descent is offset by the upward component of air movement, so that the bird remains at a uniform vertical elevation with respect to the earth. Slope soaring over a cliff face is a common practice of migrating hawks, swifts and swallows, but this involves a more complex air structure which requires more variable flight behavior than static soaring. Slope soaring is also employed by people who hang-glide. Finally, petrels and albatrosses use slope soaring over ocean waves to prey on fish.

Thermal soaring by vultures, hawks and eagles is a common sight on hot summer days. These birds utilize the energy of warm air rising from the earth's surface when the air is fairly still during mid-day. The rising warm air beneath cooler air is unstable. The warm air rises as a large vortex ring from the earth's surface (Fig. 5.5(a)), analogous to, but on a much larger scale than, the vortex ring shed from the tip of a fish's pectoral fin (Chapter 4, Fig. 4.9) or from the wing tip of a bird. The circulation of air within the thermal means that the inner air moves upward at a faster rate than the overall system. By gliding in a circular path aligned with the upward current of air in the center of the torus, large raptors are able to gain altitude with respect to the ground while descending with respect to the local air. These large birds are quite adept at moving from thermal to thermal as they hunt their prey on the ground. Various arthropods (moths and spiders) also probably use thermals to balloon themselves via an extruded length of silk as a dispersal mechanism (Vogel 1994).

Dynamic soaring involves the use of energy available in the velocity gradient of air due to wind shear over the earth's surface. At the surface, the velocity is zero (due to the no-slip condition) but increases with the square of altitude. Dynamic soaring is favored by an open expanse with a steady strong wind, conditions commonly found over the ocean. Albatrosses take advantage of this velocity gradient to oscillate in a spiral flight path (Fig. 5.5(b)), descending downwind (or at some cross-wind angle) to gain speed (and kinetic energy) before turning, as they near the ocean's surface, to fly upwind or to maneuver for feeding using the kinetic energy gained during their descent. The low wind

(a)

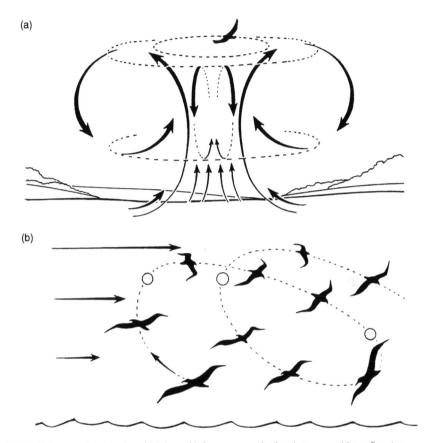

(b)

Fig. 5.5 Two mechanisms by which large birds use energy in the air to soar without flapping. (a) Thermal soaring. A bird (drawn slightly larger than life) can, by circling within a rising vortex ring, descend with respect to the local air but remain within the ring and ascend with it. (b) Dynamic soaring in a wind whose horizontal speed increases with altitude. The bird alternately ascends and descends, extracting energy from the gradient. The bird reverses its heading at the marked points. (Reproduced from Vogel (1994), Figs 10.10 and 11.15, with permission from Princeton University Press.)

velocity near the surface allows the bird to reduce its drag as it maneuvers or when it begins to fly upwind. As the bird flies upwind and begins its ascent, it not only exchanges kinetic energy for potential energy, but, by encountering increasingly faster moving air, gains additional altitude. Once it regains sufficient altitude, the albatross then turns and begins another downwind descent. Dynamic soaring allows albatrosses and petrels to travel long distances and maneuver at much lower flight costs than if they relied on powered flapping flight.

5.4 Flapping flight

5.4.1 Kinematics

Insects, birds and bats have all evolved effective powered flight by means of airfoils that produce both lift and thrust by oscillating their wings relative to their flight path (Fig. 5.6). Whereas the propellers (or jet engines) and fixed wings of aircraft carry out these functions separately, the wings of flying animals

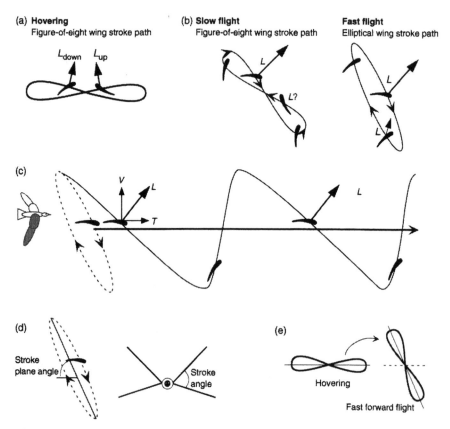

Fig. 5.6 Kinematics of wing motion and angle of attack (a) during hovering in insects and hummingbirds, and (b) during slow and fast flight in birds and bats. (c) Wing path during forward flight as a combination of the bird's forward velocity and the wing's motion relative to the bird's body. The asymmetry of the path results from the relative upstroke and downstroke motions during forward flight. The net orientation of incident air flow relative to the wing during the downstroke ensures that generation of lift L includes a component of thrust T to overcome drag D on the bird's body and wings. (d) Definitions of stroke plane angle and stroke angle. (e) Shift in stroke plane angle with change in speed. When hovering stroke plane angle is nearly horizontal (net thrust is zero). In order to fly forward, bees and hummingbirds reduce their body pitch and adjust their stroke plane angle to produce thrust as a component of lift.

must do both. Because of this the kinematics of wing movement during flapping flight are fairly complex. The wing beat cycle is most basically divided into two phases: the downstroke and the upstroke. In vertebrates, the downstroke typically produces most of the lift and thrust (as a component of lift) required for flight. During the downstroke the wing is usually fully extended to maximize wing area. By rotating the wing to reverse its orientation, insects and hummingbirds can produce additional lift and thrust during the upstroke (Fig. 5.6(a)). This allows these animals to hover for long periods. Unlike hummingbirds, which have evolved a unique shoulder articulation which allows them to rotate the wing to achieve a positive angle of attack during the upstroke (effectively a backstroke during hovering), the shoulder articulation of other birds and bats prevents substantial wing rotation. Consequently, most birds and bats flex the wing during the upstroke, particularly at slow speeds, to avoid unwanted drag and 'negative lift'. Even so, some birds and bats may achieve sufficient wing rotation to generate additional useful lift during the upstroke of slow flight. At fast flight speeds, birds with high-AR wings are generally able to sustain useful lift throughout the upstroke and downstroke, utilizing a 'continuous vortex gait' (see section 5.4.2).

Associated with airfoil rotation, the wings of many insects, hummingbirds and other birds often make a figure-of-eight pattern relative to the wing's hinge axis on the body, oscillating with a **stroke plane** that is angled relative to the body (Fig. 5.6(b)). This is not to be confused with the **stroke angle**, which represents the angle through which the wing moves during each half cycle (Fig. 5.6(d)) and determines the amplitude of the wing beat. At moderate to fast flying speeds, wing rotation during the upstroke does not occur in birds and bats, and the figure-of-eight pattern changes to an elliptical stroke movement of the wing (Fig. 5.6(b)). In birds the wing generally moves down along a path in front of its path during the upstroke, whereas in bats the elliptical path of the wing is reversed, with the downstroke passing slightly behind the upstroke path. The fact that the stroke planes of the wings are inclined, so that the wings are brought forward as they are swung down, may seem counterproductive to generating thrust for forward flight. However, when the animal's forward motion and wing rotation are taken into account, the trajectory of the wing and its angle of attack with respect to the resultant vector of the oncoming air are effectively oriented for generating thrust as a component of lift during the downstroke (Fig. 5.6(c)). The resulting motion of the wing relative to the air (its 'profile path') is asymmetrical. Its slope during the downstroke is much less steep than during the upstroke. This results from the reversed direction of the wing's movement with respect to the motion of the bird during each phase of the cycle.

In birds and bats, this asymmetry also reflects the relative timing of upstroke and downstroke, which occurs with approximately a 1:2 ratio of time for each phase (i.e. the downstroke lasts about two-thirds of the total cycle). However, this ratio can vary with speed and between species. Cockatiels have a ratio close to 1:2 at slow speeds, but approach 1:1 as speed increases (Hedrick *et al.* 2002). The air speed v_r of the wing during the downstroke is very high because the wing's own velocity v_f sums with the animal's air speed v. As a result, aerodynamic lift is high and angled forward to provide a component of thrust. During the upstroke, aerodynamic forces are much lower because the wing is flexed and wing area is greatly reduced. In addition, the feathers of birds can also rotate during the upstroke to lower their profile drag. During the downstroke, the feathers rotate back to form an interlocking array which prevents air from passing through and achieves an effective airfoil shape. This mechanism of drag reduction is unavailable to bats and insects with their solid wings. However, as we have noted, rotation of their blade-like wings allows many insects to generate useful lift during the upstroke as well as the downstroke.

In general, both the stroke plane and stroke amplitude of the wing are adjusted, together with its angle of attack, over a range of flight speeds in order to adjust the amount of lift relative to thrust that is produced. For example, during hovering, the wings of bumblebees and hummingbirds oscillate in an almost horizontal plane (Fig. 5.6(a)). To move forward at faster speeds, the animals simply incline the stroke angle of their wings to produce increased thrust (Fig. 5.6(e)). Much of the change in stroke angle relative to the horizontal is, in fact, achieved by changes in body pitch. In general, animals have a high body pitch angle while hovering and moving at slow speeds, and reduce their body pitch at faster speeds. This also helps to reduce parasite drag by decreasing the body's profile area, countering the increase due to speed (Fig. 5.3). By reducing drag, a reduction in the wing's angle of attack may provide an additional mechanism for achieving faster flight speeds.

5.4.2 Flight gaits

The steady flight of birds and bats probably involves two distinct 'gaits' (Rayner *et al.* 1986; Spedding 1987) based on differences in the vortex structure of the animal's wake (Fig. 5.7). This has been determined by flow visualization studies of air movement tracked by the motion of thousands of small neutrally buoyant helium bubbles through which the bird or bat flies. During slow flight, separate vortex rings are shed ('vortex-ring gait') at the end of each downstroke (Fig. 5.7(a)). This results from the dissipation of circulation as the wing slows down and must reverse direction. Consequently, use of this gait requires that circulation be redeveloped about the wing at the initiation of each downstroke. At moderate to fast flight speeds, however, the vorticity may remain intact as a vortex tube that is shed back from the wing tip, rather than being dissipated and separated as the wing reverses direction (Fig. 5.7(b)). This pattern is described as a 'continuous-vortex gait'. The continuity of the vortex shed

(a)

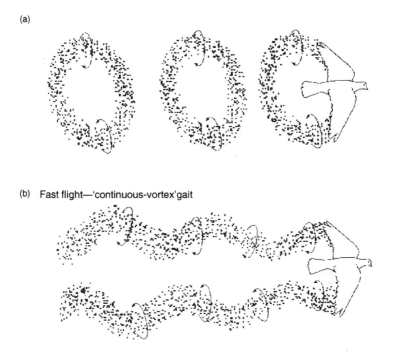

(b)   Fast flight—'continuous-vortex'gait

Fig. 5.7 Aerodynamic flight gaits used by birds: (a) vortex-ring gait used at slow speeds (and at very fast speeds in some birds); (b) continuous-vortex gait used at fast speeds. See text for details. (After Tobalske and Dial (1996).)

from the wing tip is facilitated by the reduced amplitude of the wing's motion, which can occur at faster flight speeds because of the decreased induced power requirement (Fig. 5.3). Maintenance of more uniform circulation about the wing throughout the wing beat cycle is also facilitated by moderate wing flexion, particularly at the wrist, which orients the shed wing-tip vortex more in-line with the animal's flight path and reduces drag during the upstroke. Finally, differences in wing shape are also important to the gait that a bird uses over a given range of speed. Birds with stubby low-AR wings probably use a discontinuous ring-vortex gait over their full range of speed, whereas birds with pointed high-AR wings are able to employ a continuous-vortex gait over a broader range of intermediate to fast speeds. Recent studies of cockatiels and ringed-turtle doves (Hedrick *et al.* 2002) indicate that both species change from a discontinuous vortex-ring gait to a continuous-vortex gait at about 7 m/s. However, black-billed magpies appear to utilize a vortex-ring gait over their full speed range.

In contrast with the discontinuous kinematic and mechanical nature of terrestrial gait changes, changes in aerodynamic gait during flight involve more subtle and gradual shifts in the wing kinematics and circulation patterns that they engender (Hedrick *et al.* 2002). These shifts occur as a result of the continuous change in flight velocity, which underlies whether or not the wing can maintain continuous circulation throughout the downstroke and upstroke. Interestingly, at their very fastest flight speeds, cockatiels and doves appear to revert back to the use of a vortex-ring gait, similar to that used at slow flight speeds. This flight gait is also used when these two species accelerate. The reason for this shift in gait at high speeds appears to be linked to the increasing need to reduce drag. Otherwise, the bird cannot generate sufficient thrust to move at very high speed. This suggests that a bird's top flight speed may be determined by its ability to limit the rapid increase in drag, while at the same time maintaining sufficient thrust. Shifting to a vortex-ring gait reduces the drag that the wing would otherwise experience during the upstroke if it continued to use a continuous-vortex gait. The loss in upstroke lift is not a limiting constraint because weight support is not the key problem at high speeds.

5.4.3 Bounding and undulatory flight

In addition to changing gait, many species of birds also vary their flight behavior, alternating periods of flapping flight with periods of bounding or gliding flight (Rayner 1985; Tobalske and Dial 1994; Tobalske 1996). This results in an undulatory flight path (Fig. 5.8), referred to as flap-bounding or flap-gliding (the latter is commonly distinguished from undulatory flight, but both behaviors result in oscillating upward and downward flight paths). Both appear to reflect strategies for reducing energy expenditure and improving muscle performance. In flap-bounding, the bounding phase occurs with the wings drawn in close to the animal's body. This eliminates profile and induced drag, so that the animal's body flies through the air as a projectile for brief periods. During the bound phase the animal loses altitude, which is regained during the flapping phase. The relative time spent bounding versus flapping depends on the overall speed of the animal, its size and the wing geometry. The energy savings by reduced drag during the bounding phase must exceed the net energy that is lost and must be regained during the flapping phase for the bird to achieve a net benefit. Otherwise, steady flapping flight is favored. This is most likely to be the case at faster flight speeds, when profile power is high if the wings are extended. Flap-bounding is more commonly observed in small birds (swallows, finches, warblers etc.), but medium-sized birds (magpies and woodpeckers) also flap-bound (Tobalske and Dial 1996), with a glide-bound often preceding landing to a perch. Another potential advantage is that flap-bounding allows the bird to use a constant wing beat frequency and wing stroke amplitude during the flapping phases, which may allow their flight muscles to operate at maximum efficiency (see Chapter 2,

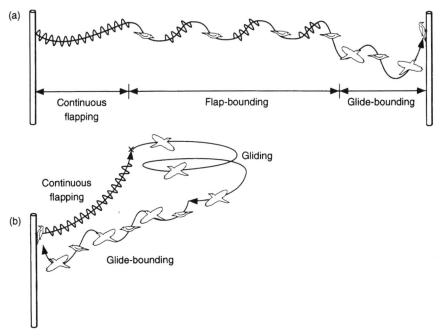

Fig. 5.8 (a) Intermittent flight behavior in birds: flap-bounding, glide-bounding and gliding flight are frequently used in association with flapping flight by different species. (b) Foraging behavior of Lewis's woodpecker. (After Tobalske (1996).)

Fig. 2.4). This 'fixed-gear' hypothesis (Rayner 1985) depends on the fiber characteristics of the pectoralis muscle being uniform (see section 5.5.1). According to this hypothesis, the bird varies the period of time that it flaps relative to bounding flight in order to operate its muscle fibers at a uniform contraction rate (e.g. $0.3V_{max}$) which maximizes their efficiency for converting metabolic energy into mechanical work, despite changes in flight speed. Nevertheless, recent work on zebra finches (Tobalske *et al.* 1999) shows that the angular velocity of the wing increases with increasing speed during flap-bounding. This indicates that the bird's flight muscles must also increase their contraction speed, which would limit their ability to operate as a simple 'fixed gear'.

Flap-gliding involves intermittent gliding periods interposed between flapping periods when the animal must regain the kinetic and potential energy lost during the glide. Here again, the energy savings during the glide must offset the energy cost to regain altitude and speed during the flapping phase. In the case of flap-gliding, this can occur at low to moderate flight speeds because gliding is effective for generating lift over this speed range and profile power is low. The particular flight path depends on the climb angle, the subsequent glide angle and the relative time spent gliding. Although flap-gliding is commonly considered to be a characteristic behavior of larger birds and bats with low wing loading and high-AR wings, which favors gliding performance, many exceptions exist. For example, crows and jays with intermediate wing loading and low-AR wings, as well as much smaller swallows with high-AR wings, all regularly flap-glide. Nevertheless, birds with very high wing loading, such as ducks, do not.

5.4.4 Origin and evolution of flapping flight

Various theories have been advanced for the evolution of flight in animals. The role of a wing as an airfoil for flight is obvious, but its use for other functions is less clear. Specifically, it is important to determine what selective forces would favor intermediate stages in the evolution of such a device and the means by which, in the case of birds, a fully-fledged flight feather may have evolved. The evolution of feathers for insulation or display represent two reasonable hypotheses. The problem of intermediate design is common in evolutionary biology, and the evolution of wings is a classical example. Two principal theories have been proposed for the evolution of flight in vertebrates: the 'trees-down' theory and the 'ground-up' theory (reviewed by Norberg (1990)). The first depends on the ability of proto-fliers to climb, the advantages of being arboreal and, finally, an advantage of developing a lift-producing airfoil. There is evidence of free claws on the hands of early birds which, since they were small, would also favor their ability to climb. The earliest pterosaurs, like bats, were also small and probably used their forelimb claws for climbing. The advantages of being arboreal probably were (and are) safety from predators and the availability of new foraging sources. Given the ubiquitous forms of other gliding and parachuting arboreal vertebrates, the selective advantage associated with evolving an extended body surface seems clear. A further increase in such a surface would be favored because it would improve the animal's gliding performance. Gliding also takes advantage of gravity and provides a clear energetic savings for the animal in terms of foraging cost. Once evolved as an effective gliding airfoil, the evolution of flapping wings for powered flight is not difficult to envisage.

The above scenario appears attractive. However, there is no evidence that any living obligate gliding species flaps its gliding surface to improve its glide performance. Gliding on its own is a successful means of escape or economical movement through an arboreal habitat. Finally, when models of gliding airfoils are flapped, they usually have reduced lift performance.

The ground-up theory requires that incipient wings were either used for prey capture (as in the case of birds) or for improved stability when maneuvering or running fast. However, these scenarios also present some problems. Modern ground-dwelling birds do not use their wings as airfoils to stabilize themselves when running and maneuvering. Dynamic stabilization of the body's center of gravity over the supporting limbs is effectively accomplished without the use of the forelimbs (see Chapter 3, section 3.6). In addition, extended wings would increase drag and slow the animal down, reducing its ability to capture flying insect prey or to escape a pursuing predator. Finally, the ground-up theory requires that any incipient flight capability would have had to overcome gravity. Not only would the proto-flier have had to run fast in order to reduce its induced power cost for take-off, it would have had to do work against gravity in order to achieve any energetic advantage from gliding.

A more recently proposed theory is that flight may have evolved in association with the use of incipient wings to assist a ground-dwelling bird to climb up steep slopes (Dial 2003). Ground birds prefer to be in elevated locations when these are available as this reduces their predation risk. Dial's theory is based on studies of the ontogeny of climbing and wing-flapping behavior in galliforms. Unlike when they run on a level surface, galliforms flap their wings when they climb up an incline. Juvenile birds with short wings have greater difficulty in climbing steep slopes. With the development of the wing and feathers, larger older birds are able to climb better. Dial's (2003) hypothesis is that the evolution of flight may have involved a similar selection pressure for wing-assisted climbing, much like the pattern observed in the ontogeny of locomotor performance in galliforms. The exact mechanism of how the wings assist the bird to climb (either by helping it to remain in contact with the ground for better grip, or by providing a more upward force to assist its climb) is not yet known.

The evolution of flight in insects has been argued along three main theoretical lines. One theory (Wasserthal 1975; Douglas 1981; Kingsolver and Koehl 1994) posits that insect wings originally evolved as thermoregulatory devices which allowed the animal to regulate heat loss and gain (by solar radiation) through changes in surface area and wing orientation. Once they had evolved as thermoregulatory structures, selective advantages for gliding performance and ultimately powered flapping flight, similar to those described above for birds and bats, could have been realized. A second possibility, similar to the ground-up theory for birds, is that proto-wings evolved as extensions from the legs or from gill-like appendages in jumping insects. Any increase in aerial performance during the jump (to escape predation) would presumably have been selectively advantageous. A third theory (Marden and Kramer 1994), introduced in Chapter 4, section 4.8.1, argues that wings may have evolved initially as sails, allowing insects (modern analogs being stoneflies) to sail over the water surface, which would presumably have provided dispersal and foraging benefits.

All three theories seem plausible. However, current evidence indicates that insect flight arose once in a common ancestral proto-flier (Dudley 2000), so that only one of these, or some other, scenario actually occurred. Heinrich (1993) points out that the advantage of lateral lobes (proto-wings) for more rapid heating is of little value to an animal as small as an insect. Without their wings, butterflies still heat up at impressive rates (25 °C/min) (Heinrich 1993). Heinrich also notes that endothermy and thermoregulation are *only* associated with insects that fly. No known living insect basks or shivers to heat up, except just before flight. An attractive aspect of the sailing hypothesis for the origin of insect flight is that fossil insects possessed gills or gill covers, capable of being moved for ventilation of the water, which are quite reminiscent of the reduced wings used for sailing.

As for vertebrates, the evolution of flight in insects requires a plausible hypothesis for selective advantages associated with incipient airfoils and their initial and intermediate function. These requirements are met by both the thermoregulatory and sailing models of insect flight evolution, in which any increase in surface area would have improved functional capacity. Subsequent modification of the wing's shape as an airfoil and internal modifications of the musculature and skeleton for active generation of aerodynamic lift would have been favored by any initial benefits achieved by means of improved gliding performance.

5.5 Flight motors and wing anatomy

Because flapping flight requires considerable power, the flight muscles of vertebrates and insects achieve some of the highest capacities for sustained mechanical power within the animal kingdom (in the range of 200–400 W/kg muscle). These muscles are generally organized as antagonist groups which either depress or elevate the wing. Because wing depression generates most of the aerodynamic lift in bird and bat flight, the wing depressors are considerably larger muscles than the wing elevators. In contrast, with aerodynamic lift produced during both the upstroke and downstroke of most flying insects, their flight muscle antagonists are more similar in size. Associated with their ability to generate significant lift during the upstroke, hummingbirds similarly tend to be distinguished by having relatively larger wing elevators than other birds for their size. Whereas the flight muscles of flying insects operate at very high frequencies and contract over only a small fraction of their length, the flight muscles of larger birds operate at lower frequencies but contract over much greater ranges of length.

In addition to the muscle machinery that powers flight, many other muscles control wing orientation, wing shape and adjust the stroke plane. These muscles, referred to as steering muscles in insects, are important to the maneuvering flight of birds and bats as well as insects. Less is known about these muscles, which are more numerous and smaller in size. Our focus here will be on the larger flight muscles that are most important to lift generation. In addition to the flight musculature, the design of the skeleton is also important to how the muscles transmit their force to the wing. Finally, as we have discussed in some detail, the design of the wing itself is critically important to its function as an airfoil.

5.5.1 Vertebrate flight musculature

In birds and bats, the pectoralis is the primary muscle which depresses the wing in order to produce lift. The pectoralis originates from the body via the sternum, ribs and clavicles (in birds the clavicles are fused to form the furcula or 'wishbone') and attaches to the humerus, the most proximal bone in the wing (Fig. 5.9). In birds, the pectoralis muscles together constitute 12–22 per cent of the total body mass. In many species, the pectoralis attaches locally to a bony process which projects anteriorly from the humerus, termed the deltopectoral crest. Its anterior insertion means that the pectoralis also tends to rotate the wing in a nose-down direction (pronation). This is necessary to balance the opposing moment produced by aerodynamic force on the wing,

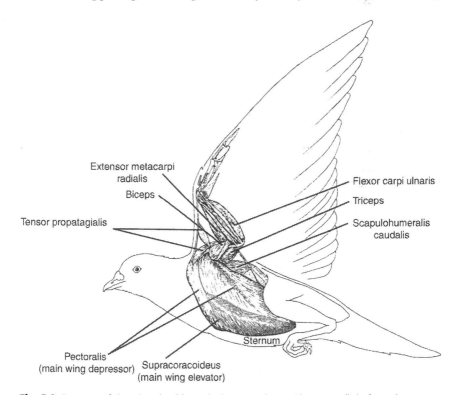

Fig. 5.9 Anatomy of the avian shoulder and wing musculature. The pectoralis is the main downstroke muscle, and the supracoracoideus (shaded) which lies deep to the pectoralis elevates the wing (Reproduced from Dial (1992a), with permission Wiley-Liss, Inc.)

which acts distal but anterior to the elbow. As a result, torsional loading of the humerus is common in the flight of bats (Swartz *et al.* 1992) and birds (Biewener and Dial 1995). In birds, the main elevating muscle of the wing, the supracoracoideus (homologous to the pectoralis minor of bats and other mammals), has an unusual and intriguing anatomical arrangement. It also arises from the keel of the sternum deep to the pectoralis. Consequently, it has the same ventral position as the pectoralis relative to the wing. However, the supracoracoideus elevates the wing by means of a pulley-like arrangement of its long tendon, which passes anterior to and over the shoulder to attach to the dorsal aspect of the humerus. In most birds the supracoracoideus is about one-tenth of the size of the pectoralis; however, in hummingbirds it is about half the size. This difference is consistent with the fact that lift is generated during the upstroke as well as the downstroke in hummingbirds, but is believed to be passive in most other birds. In bats, both the pectoralis major and minor act as depressors of the humerus at the shoulder; consequently, wing elevation is mainly achieved by the deltoid and other shoulder muscles, which insert on to the dorsal aspect of the humerus.

The pectoralis and supracoracoideus are pinnate muscles in birds, enabling them to generate large forces for their mass. The pectoralis has much longer fibers than the supracoracoideus, associated with its larger moment arm and need to produce a large ventrally directed torque for generating mechanical power during the downstroke. The fiber type characteristics of the pectoralis muscles of several avian species have been examined (George and Berger 1966; Rosser and George 1986). In general, the pectoralis of flying birds consists largely of fast-twitch muscle fibers. The majority (80–90 per cent) can be characterized as being fast oxidative glycolytic (FOG) (see Chapter 2, section 2.1.6), with most of the remaining being fast glycolytic (FG). Because flapping flight generally requires sustained high-frequency contractions, the slow-oxidative (SO) fibers found in the muscles of terrestrial animals and fish are largely absent from the flight muscles of many birds. However, in soaring birds, such as turkey vultures and frigate birds, specialized groups of SO twitch and/or slow tonic fibers have been identified. These fibers enable economical and sustained isometric contraction of the pectoralis associated with prolonged gliding flight. The uniformity of fiber type characteristics within the flight muscles of avian species is consistent with the observation that wing beat frequency and angular velocity change little over a range of flight speed and flight mode (Dial and Biewener 1993; Tobalske and Dial 1996; Dial *et al.* 1997). Hence, having flight muscles which operate at a uniform contractile speed appears to be a general feature of birds and is not necessarily limited to smaller species that use bounding flight. FOG and FG fibers similarly predominate in the flight muscles of bats; however, greater variation appears to exist among bats, including more prevalent populations of SO fibers (reviewed by Norberg (1990)). The heterogeneity of histochemically defined fiber types across diverse taxa makes their classification and the inference of their physiological properties problematic. Consequently, more work correlating muscle physiology with fiber-type characteristics is needed for understanding how differences in flight muscle design are correlated with flight capabilities.

Other muscles contribute to motions of the wing in birds, but they are mainly used to control wing shape and airfoil orientation important to maneuvering and control of landing (Dial 1992b). This is also the case for bats, where changes in wing shape by differential activation of several shoulder and forelimb muscles allow them to achieve highly maneuverable flight (Altenbach and Hermanson 1987). Finally, it is likely that much of wing elevation may be passively achieved by aerodynamic lift during steady forward flight at moderate to fast speeds. This means that the pectoralis represents the primary power-generating muscle in birds and bats.

5.5.2   Avian pectoralis function: implications for power output during flight

The unique anatomy of the pectoralis and its insertion on the humerus of certain birds allows direct recordings of the forces and length changes produced by the pectoralis during flight (Dial and Biewener 1993; Biewener *et al.* 1998b). These recordings show that the pectoralis generates work loops (Fig. 5.10) similar to those developed by fish axial musculature and bumblebee flight muscles (see below). In order to perform work, the pectoralis lengthens only slightly, if at all, late in the upstroke (A), allowing it to develop considerable force (B) before it shortens during the downstroke to do aerodynamic work (C). The muscle then relaxes, allowing it to be passively lengthened during the upstroke (D). The main difference between fish axial muscle work loops and those of the avian pectoralis is that the latter undergoes substantial length changes (30–40 per cent of resting length), associated with its need to move the wing through a large angular excursion.

Similar measurements of pectoralis force and kinematic estimates of muscle length change in black-billed magpies flying in a wind-tunnel (Dial *et al.* 1997) indicate that their power requirements for flight differ substantially from the U-shaped curve predicted by classical aerodynamic theory (Fig. 5.11(a)). These measurements show that, whereas the power requirement is highest during hovering flight, it rapidly declines by two- to threefold as the bird increases its forward speed and remains fairly uniform over a range of speeds from 4 to 14 m/s. A slight increase in power output due to increases in profile and parasite drag is observed at the highest flight speeds. Over the full range of speeds wing beat frequency changes very little. The ability of magpies to fly with fairly uniform mechanical power over a range of speeds is probably due

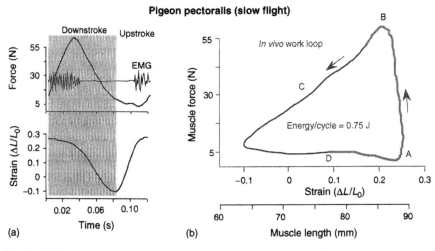

Fig. 5.10 (a) Force, EMG and length change (strain) of the pigeon pectoralis over one wing beat cycle. (b) The work loop (force versus length) produced by the muscle to power the bird's flight. The area within the loop represents the net positive work performed over one contraction cycle. Arrows denote the path of muscle force relative to length change versus time. The light shading denotes the time during which the muscle is activated based on its EMG.

to changes in flight behavior (relative phase of flap-gliding sequences) and wing shape (Tobalske and Dial 1996). Part of the discrepancy between the direct measurements of power and theoretical predictions is the suggestion by aerodynamic theory that birds, such as magpies, should be able to sustain flight speeds at a power that equals or exceeds that required during hovering flight (see Fig. 5.3). However, for most birds the power required for hovering, if they ever attempt to do it, probably requires anaerobic sources of energy supply to the muscles and hence is non-sustainable. Consequently, birds are unlikely to use faster flight speeds which require comparable non-sustainable power in the course of their normal flight behavior.

Recent studies of cockatiels and ringed turtle-doves flying in a wind tunnel (Tobalske *et al.* 2003) however indicate a clearly U-shaped power curve (Fig. 5.11(b)). Measurements of metabolic cost versus flight speed in parakeets (budgerigars) also indicate a U-shaped power curve. It would be interesting to know whether the metabolic power requirements of cockatiels and ringed turtle-doves reflect their mechanical and aerodynamic U-shaped pattern. However, metabolic studies of other species indicate much flatter power curves (see Chapter 9, Fig. 9.13). Metabolic measurements are typically limited to a fairly narrow range of flight speed because they require much longer periods of flight in order to obtain reliable (steady state) aerobic measurements. Mechanical measurements of muscle power can be made over

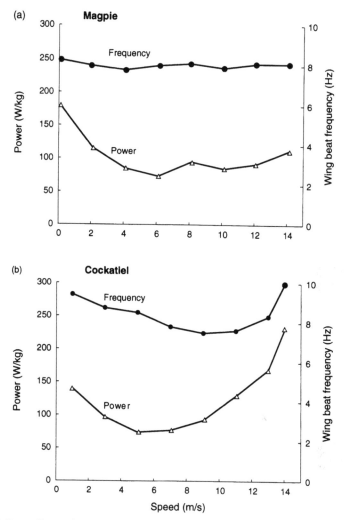

Fig. 5.11 Pectoralis-muscle-specific power output and wing beat frequency versus speed for (a) magpies (Dial *et al.* 1997) and (b) cockatiels (Tobalske *et al.* 2003).

a smaller number of wing beats for both aerobic and anaerobic non-sustainable flight. Consequently, except for hummingbirds (and insects), metabolic measurements of flight cost are almost impossible to make during hovering and slow-speed flight and are also difficult to obtain for very fast flight. Future studies combining aerodynamic, muscle power and metabolic power approaches are needed to assess the degree to which variation in a classical U-shaped power curve exists among different species which vary in body size, wing and tail geometry and flight behavior. It seems likely that at least some species will diverge from the standard power requirements derived from steady state aerodynamic theory.

The results obtained for muscle power in magpies, pigeons (Dial and Biewener 1993), starlings (Biewener *et al.* 1992) and, most recently, cockatiels indicate that the pectoralis muscle operates with an efficiency in the range 15–23 per cent. This is consistent with the maximum efficiency of vertebrate skeletal muscle being in the range 20–25 per cent.

5.5.3 Insect flight muscle mechanics

In general, the flight muscles of insects consist of direct muscles which attach from the thorax to the wings and indirect flight muscles which lie within the thorax but do not attach to the wing. The contractile properties of these muscles are distinguished among various insect fliers by being either synchronously or asynchronously activated. As their name suggests, synchronous muscles (found in locusts, beetles, moths, dragonflies and other large insects) contract in a 1:1 ratio with respect to the firing frequency of their motor nerves, limiting their frequency in most species to less than 150–200 Hz. In contrast, asynchronous flight muscles (found in bees, wasps and flies) operate at frequencies considerably above 200 Hz, ranging as high as 1000 Hz (the wing beat frequency of mosquitoes, for example, is about 500 Hz, which gives them their telltale and irritating 'hum' as they fly close to one's ear). Asynchronous muscles are not directly activated by their motor nerves. Instead, their high contractile frequency is achieved by means of being stretch-activated by their antagonist in combination with having to contract against the inertial mass of the wing. This distinction is mainly based on the physiological properties of the muscles, rather than their anatomical organization. When a nerve stimulus is transmitted to asynchronous muscles, they contract many times rather than once.

Synchronous and asynchronous flight muscles can generally be divided into two antagonist groups: wing elevators and wing depressors. This is an oversimplification because certain muscles within these groups also rotate the wing as it swings back and forth in a figure-of-eight loop, as well as protract and retract the wing between bouts of flight and rest. The relative size and organization of the direct and indirect flight muscles, in part, reflects whether they are synchronous or asynchronous. The direct flight muscles of synchronous insects are generally larger than the indirect flight muscles, whereas in asynchronous fliers the indirect flight muscles have evolved to become the dominant muscles powering flight. Two sets of indirect flight muscles are identified: a dorsoventral pair and a dorsal longitudinal pair (Fig. 5.12). Their alternating contractions deform the thorax, causing the wings to be either elevated or depressed. Though still not well understood, changes in thorax shape produced by alternating contractions of these muscles induce wing rotation at the wing hinge.

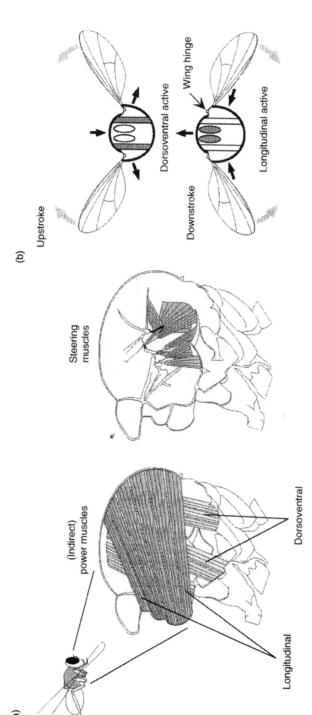

Fig. 5.12 Insect flight muscle organization. (a) Anatomical drawing showing the indirect muscles which power flight and the smaller steering muscles which control wing orientation and rotation. (b) Schematic model of the insect thorax, showing how the dorsoventral and longitudinal asynchronous flight muscles indirectly elevate and depress the wings via deformations of the thorax (bold arrows). (Courtesy of M. Dickinson.)

The elastic stiffness of the thorax relative to the underlying contractions of the flight muscles is important to its operation as a high-frequency resonant system. Contraction of the dorsoventral muscles causes the thorax to compress dorsoventrally and expand longitudinally and laterally, elevating the wings. This also stretches the longitudinal muscles, causing them to contract, which compresses the thorax longitudinally and expands it dorsoventrally, producing wing depression. Their contraction, in turn, stretches and activates the dorsoventral muscles, causing the cycle to be repeated. Activation by the motor nerves is believed to be important for both the initiation of flight activity and the maintenance of muscle tone needed to sustain the high-frequency stretch-activated sequence of muscle contractions. Once motor nerve impulses cease, the intrinsic operation of the flight muscles quickly ends.

In contrast with the large strains that the pectoralis of birds undergoes to power the motions of the wing, the main flight muscles of insects operate at much lower strains (<5 per cent of resting length). The low strains of asynchronous flight muscle are consistent with its highly structured organization (Chapter 2, Fig. 2.1) and its need to operate at a very high frequency. By operating at a high frequency, asynchronous flight muscles achieve a power output in the range of 100 to 200 W/kg muscle, similar to the range for the pectoralis of birds. Synchronous muscles achieve power outputs that are approximately half that of asynchronous flight muscle, consistent with the view that the evolution of asynchronous muscle enabled an increase in power output and flight performance via high contraction frequencies (Josephson et al. 2000). Roughly 75 per cent of flying insects have evolved asynchronous flight muscles (Dudley 2000). The efficiency of invertebrate flight muscle has been determined to be in the range 9–15 per cent. The lower range compared with vertebrate muscles probably reflects limits imposed by the need to operate at a high contractile frequency.

5.5.4 Insect thermoregulation in relation to flight

In order to operate at high frequencies, the muscles of flying insects must be warmed before flight. This allows their flight muscles to be activated, develop force and relax at much higher rates than would otherwise be possible at lower temperatures. Flight warm-up is a characteristic thermoregulatory behavior of most flying insects, particularly those that inhabit more temperate climates and those that are nocturnal and must fly during cooler night-time temperatures. For the interested reader, Heinrich (1993) describes in detail the physiology and related flight ecology of 'hot-blooded' flying insects in his delightful and clearly written book. Many moths, butterflies and dragonflies either utilize flight muscle shivering thermogenesis or basking and solar heating to warm their thorax to temperatures of 32–40 °C prior to flight. Basking is a common strategy for diurnally active moths, butterflies and dragonflies, whereas muscle thermogenesis is needed to heat up at night or early in the day. Solar heating is an obvious function of the large wings of these species.

Insect body size also affects flight performance because of its effect on heating and cooling. Although larger moths and butterflies require a longer period of time to warm up, once they reach an appropriate flight temperature, they can sustain continuous flight for longer periods than smaller species. Smaller species tend to cool when they fly owing to convective heat loss, which limits their flight to briefer durations. Heinrich (1993) demonstrated that moths thermoregulate during continuous flight by pumping warmed blood which passes through their flight muscles via the dorsal aorta to their abdomen where any excess heat produced by the flight muscles is lost to the air. The abdomen is relatively uninsulated and thus acts as a radiator in combination with the heart to control rates of heat loss and maintain thoracic temperatures from becoming dangerously high. During pre-flight warm-up, blood flow to the abdomen is reduced so that convective heat loss is minimized at a time when the moth seeks to elevate its thoracic temperature as rapidly and economically as possible.

In contrast with tropical bees, temperate bumblebees and honeybees also exhibit pre-flight warm-up to elevate their thoracic temperature. To do this, they contract their flight muscles synchonously via direct neurogenic stimulation. Once a sufficient thoracic temperature is reached (typically 30 °C or more), the flight muscles switch to asynchronous stimulation allowing them to achieve the high wing beat frequencies necessary for flight. Of all the flying insects, bumblebees exhibit the finest control and greatest capacity for thermoregulation during flight, utilizing mechanisms of heart and abdominal blood shunting control of heat loss to balance muscle heat production similar to that of moths and butterflies. In general, Heinrich argues that the evolution of thermoregulation is probably linked to large body size, at which the danger of overheating (above 45 °C) during flight due to excessive heat production by the flight muscles becomes a problem.

While many lepidopterans, dragonflies, bees and flies utilize pre-flight warm-up and thermoregulate during flight, many smaller flying insects do not and are able to fly with thoracic temperatures as low as 0 °C. Clearly, the capacity to fly with low muscle temperatures required selection for much faster myosin-ATPase and metabolic enzyme rates, as well as more rapid calcium ion release and uptake by the sarcoplasmic reticulum, and neural properties that enable fast contractile rates at low temperatures. Cold-adapted species also exhibit other adaptations. Winter moths generally have low wing loading, reducing their aerodynamic power costs and in some species only the males fly (females do not attempt to carry their eggs). The lack of a gut in males further reduces their transport cost. Finally, less regular and shorter-duration fliers, such as beetles and grasshoppers, do not undergo pre-flight warm-up and show little evidence of temperature regulation in relation to flight.

5.6 Maneuvering during flight

Not only must flying animals achieve the means for producing sufficient aerodynamic force to support their body weight and generate thrust, they must also be able to maneuver in their aerial environment. Despite its fundamental importance to the success of flying animals, maneuverability and the mechanisms underlying maneuvering during flight have been relatively poorly studied. This is in large part due to the more challenging requirements for observing and investigating the maneuvering of a flying animal. Maneuvering is not only important for negotiating obstacles, particularly in complex spatial environments, catching prey and avoiding predators, but is also used for ritualistic display and mating. Certainly, trade-offs exist between wing designs that favor more economical lift generation but which are poor for maneuvering, such as the wings of an albatross, and those that enhance maneuvering ability. In general, smaller size and shorter more rounded wings make an animal more maneuverable. Therefore smaller flying animals can successfully avoid predatory strikes of larger flying species and can exploit a broader array of spatial environments. It is likely that maneuverability has been fundamental to the evolutionary success of flying insects and passerine birds. The aerial maneuvers of a bird, a bat or an insect represent some of the most dramatic and spectacular acrobatic feats that one can observe in animal movement.

Maneuvering during flight generally involves an asymmetry in the aerodynamic forces that are produced by the animal's wings. This force asymmetry may be produced by various mechanisms. Changes in incident air velocity over the wing, the wing's angle of attack and the surface area of the wing are three variables important to lift generation which may be modulated by a flying animal to produce an asymmetric force distribution between opposing wings, enabling it to turn, dive or ascend (Warrick and Dial 1998). In addition to initiating a turn via a force asymmetry, a flying animal must subsequently produce a counteracting force asymmetry to arrest its rolling or pitching movement in order to re-establish a stable flight trajectory. By producing greater lift force on the outside wing, or by producing greater decelerating drag force on the inside wing, an aerialist can achieve a turning moment. In a recent study of slow flight, Warrick and Dial (1998) found that the primary mechanism used by pigeons to initiate and stabilize a turn is achieved primarily by means of differential downstroke velocities of the bird's wings. In order to bank and turn to the left, a slow-flying pigeon increases the downstroke speed of its right wing to initiate a turn to the left and subsequently, in the next wing stroke, counteracts the momentum of its banking turn by increasing the downstroke velocity of its left wing to stabilize its roll and continue steadily along a new heading

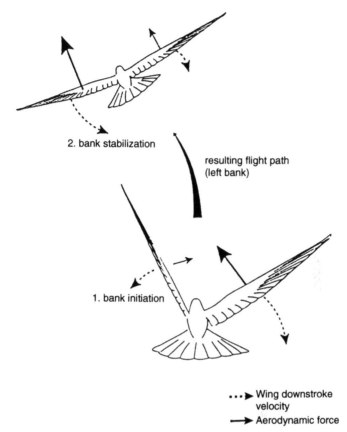

2. bank stabilization

resulting flight path
(left bank)

1. bank initiation

•••▶ Wing downstroke
velocity
——▶ Aerodynamic force

Fig. 5.13 Maneuvering left turn of a pigeon via asymmetric wing motion (dashed arrows) leading to asymmetric aerodynamic forces (dark arrows) produced by the two wings. In (1) the right wing's velocity and force is greater to initiate a leftward bank and associated body roll. This is followed in (2) by a reversed asymmetry of wing motion and force to arrest the bird's banking momentum (from Warrick and Dial (1998); permission Company of Biologists, Ltd.).

(Fig. 5.13(a)). Use of asymmetric wing stroke velocity has the advantage that lift force depends on the *square* of air velocity. Hence, use of changes in wing velocity during the downstroke provides a very effective means for creating a force imbalance to initiate and control a maneuver.

Although changes in angle of attack and wing area may not be as important at low flight speeds, this may reflect the fact that the wing's stroke velocity is high relative to the animal's forward flight speed under these conditions. At faster flight speeds and a lower induced power requirement, it may be that more subtle mechanisms, such as modulation of angle of attack and/or wing area, are used to control an animal's flight trajectory. Finally, for slower turns flying animals may simply bank their wings (Fig. 5.13(b)), so that the net orientation of lift is shifted toward the body axis by the outside wing in the direction of the banking turn. While this may be sufficient for wider-angle turns, it may not be as effective for achieving tighter turns, in which active control of force asymmetry through differences in wing velocity, angle of attack and wing shape may play a more prominent role.

In addition to these three variables, insects employ a fourth mechanism to maneuver and control their flight path. Dickinson *et al.* (1999) found that flies may use unsteady lift-generating mechanisms (see section 5.6.1) to execute steering maneuvers. By altering the timing of wing translation relative to wing rotation associated with wing stroke reversal (i.e. as the wing completes its downstroke it is rotated to enable lift generation during its reversed motion during the upstroke, and vice versa during the reversal from upstroke to downstroke), a flying insect may alter the lift that its wing generates. For example, in order to turn to the left a flying insect could *phase delay* rotation of its left (inside) wing and *phase advance* rotation of its right (outside) wing. This would result in decreased lift by its left wing and increased lift by its right wing (see below), causing it to turn left. The use of phase asymmetries in the timing of wing rotation relative to wing translation by insects may reflect the fact that airfoil shape is relatively constant compared with the more variable wing shape that birds and bats are able to use throughout a wing beat cycle. In addition, because of the indirect flight muscle apparatus that insects use for powering their flight, alteration of wing stroke velocity between opposing wings may not be a viable mechanism for achieving steering maneuvers as observed during slow flight in birds. Conversely, the use of asymmetric wing stroke phase, or timing, for maneuvering in birds and bats has not yet been observed.

5.7 Unsteady mechanisms

Most of what has been discussed in this chapter and the previous one on swimming assumes conditions of steady fluid movement over an airfoil (or hydrofoil). Steady state aerodynamic theory tells us much about the design requirements and performance of swimming and flying animals. However, it is clearly the case that unsteady flow conditions also operate in the biological realm of animal movement because propulsion is nearly always achieved by means of a reciprocating appendage (wing, tail, fin) which must be accelerated and decelerated as it reverses its direction during each half-cycle of movement. Consequently, whereas fixed-wing aircraft are well modeled and designed according to steady state flow conditions, unsteady flow conditions are likely to be important particularly during the rotational and reversible movements of an oscillating wing or hydrofoil. To emphasize this point, conventional aerodynamic theory based on fixed wings that move at a constant velocity does not explain how many species of insects are able to support their weight, let alone how they manage to carry loads (e.g. a honeybee) and maneuver. Conventional mechanisms simply do not provide enough lift for a flying insect to stay in the air (Ellington 1984). Hence, unsteady mechanisms probably operate to provide the additional lift needed for weight support and load carrying. Because flows under non-steady conditions are necessarily complex and complicated, and at present relatively little theory exists to explain such flow phenomena, a detailed consideration of unsteady aerodynamic mechanisms is largely outside of the scope of this book. Nevertheless, a brief introduction to and discussion of these phenomena is warranted, as much of the exciting and recent discovery of novel lift-generating mechanisms important to flight and maneuvering, particularly at the moderate Reynolds numbers at which insects operate, is based on unsteady flow.

A major problem limiting the performance of a reciprocating airfoil is that it must shed the vorticity developed during its previous half-stroke (e.g. downstroke) and subsequently re-accelerate the fluid moving over it to develop a new (bound) circulation in the opposite direction during the next half-stroke (e.g. upstroke). Once developed, the circulation must again be shed as the wing reverses its direction to begin the following downstroke, and so on. In the case of insects and hummingbirds, it is clear that the shedding of bound circulation and the subsequent redevelopment of new circulation must occur during each half-stroke (given that useful lift is produced during the upstroke, as well as during the downstroke, of these animals). In the case of birds during slow flight, it is less clear if significant circulation for lift is developed during the upstroke of the wing beat cycle. If not, then the problem of vortex shedding and circulation redevelopment only occurs once during an entire wing beat cycle, being partitioned at the start of the downstroke (circulation redevelopment) and at the end of downstroke (vortex shedding). Nevertheless, the need to shed vorticity and redevelop circulation as a wing reverses its direction, which is known as the **Wagner effect**, classically is considered to diminish the aerodynamic performance of a reciprocating airfoil.

The first unsteady effect to be identified and considered important to insect flight was a rotational mechanism called the **clap and fling** (Weis-Fogh 1973). This involves rapid apposition of the two wings at the end of upstroke (clap), which enhances the redevelopment of circulation as the air is sucked between the wings by their rapid rotation as they peel apart at the start of downstroke. Although important as an unsteady effect for enhancing lift in small insects and butterflies, the clap and fling is not used by all insects and so does not provide a general solution for meeting the force requirements of insect flight. A second unsteady mechanism identified more recently using a robotic model of a hawkmoth (Ellington *et al.* 1996) involves the development of a leading-edge vortex during the wing stroke which delays stall and enhances lift. This is likely achieved by enhancing the magnitude and duration of circulation developed during both the downstroke and the upstroke. The leading-edge vortex involves a spanwise flow of air from the proximal to the distal end of the wing as it beats down and up. This allows the wing to maintain circulation at a higher angle of attack than would be possible based on conventional aerodynamic flow.

In more recent experiments, using robotic model fly wings operating in a vat of mineral oil scaled to the Reynolds number regime appropriate for flies (Re = 100–200) (Fig. 5.14), Dickinson *et al.* (1999) found that insect wings probably employ two additional unsteady mechanisms: **rotational lift** and **wake recapture**. Rotational lift is achieved by the rapid angular rotation of the wing that occurs during each half stroke at wing reversal. It is similar to but more effective than the enhanced circulation that is achieved by a spinning ball as it moves through the air, which is known as the **Magnus effect**. Owing to the ball's spin, the velocity of airflow is increased on one side of the ball (the side in which the rotational velocity is counter to the ball's own motion) and reduced on the opposite side. This creates a pressure differential (lower pressure on the high-velocity side relative to the low-velocity side) which causes the ball to curve in the direction of its spin (this explains how backspin causes a ball to rise or a baseball to curve due to the spin imparted by a pitcher as it is released from his hand). In the case of a fly wing, the lift produced by the wing's rotation is considerable, amounting to as much as 35 per cent of the total lift generated by the wing. Finally, wake recapture provides additional lift by means of the wing interacting with its own wake as it reverses direction and passes back through the wake produced by its movement in the previous half-stroke. This allows the wing to hasten the development of circulation as it is re-accelerated following stroke reversal, reducing the Wagner effect. The timing of wing rotation relative to its translation has a significant effect on the amount of lift that is generated. Hence alterations of the relative phase of wing rotation during the stroke cycle provide insects with an effective means of producing lift asymmetries important to steering maneuvers.

Delayed stall, rotational lift and wake recapture represent three distinct, yet interactive, mechanisms of unsteady lift generation which are necessary for flying insects to achieve the flight forces needed to support their weight and carry loads. Conventional aerodynamic mechanisms are insufficient to achieve this. Unsteady mechanisms provide sufficient lift and account for the energy loss that occurs due to vortex shedding and the need to redevelop circulation about a reciprocating wing which also must function as the animal's propeller to generate thrust. These unsteady effects may raise the lift coefficient for *Drosophila* wings by more than a factor of 2.5 compared with that achieved during steady state conditions (Dickinson *et al.* 1999). These studies also point to the importance of being able to model the Reynolds number regime appropriate to a small insect wing operating at high frequencies. Without such physical models, which are amenable to study by virtue of being scaled up in size, the ability

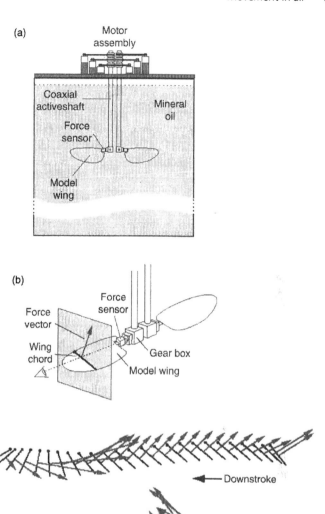

Fig. 5.14 (a) A robotic model of a fly's wings driven back and forth in a large vat of mineral oil is used to study unsteady mechanisms of lift generation. The frequency of motion and enlarged size of the model fly wings are matched to the viscosity of the mineral oil to achieve a Re appropriate for the flight of a real fly. (b) Measurements of wing's angle of attack and the forces (gray arrows) acting on the wings at successive phases of a wing beat cycle are shown below. See text for further details. (Reprinted with permission from Dickinson *et al.* 1999, Copyright American Association for the Advancement of Science.)

to measure flight forces and to image flow under unsteady dynamic conditions would not be possible. Such approaches are opening up exciting new perspectives on the complexity and beauty of fluid propulsive mechanisms used by animals. It is particularly interesting that unsteady mechanisms—wake recapture and rotational lift—can be used to control lift generation by means of the timing of wing rotation relative to wing translation, providing a mechanism for steering maneuvers as well as lift augmentation. Whether or not larger vertebrate fliers, operating at higher Reynolds numbers, make use of similar or other unsteady mechanisms to enhance the aerodynamic performance of their wing and to control their movement remains to be seen.

5.8 Summary

Flight has proven an enormously successful mode of movement, witnessed by the impressive diversity of insects, birds and bats. These are all the most speciose taxa within their respective taxonomic groups. Relative to human-engineered aircraft, the flight performance of animals is impressive. In this chapter we have reviewed fundamental features of animal wings which are important to their aerodynamic properties and the ability of animals to fly. Similar fluid-mechanical principles underlie both swimming and flying. In contrast with swimming, weight support is key to successful flight performance because of the much lower density of air compared with water. This not only limits the size range of flying animals compared with swimming animals but also means that flying animals typically move at much faster speeds. The capacity for flight based on the mechanical power derived from striated muscle is impressive, particularly given the remarkable maneuvering ability of most flying animals. The design and contractile performance of flight muscles are now being linked to the aerodynamic performance of the wing.

Because flying animals can manipulate the shape and orientation of their wings, they have the ability to utilize unsteady aerodynamic mechanisms to generate lift. This provides them with enhanced performance which exceeds that predicted by analyses of fixed-wing aircraft. However, this also provides greater challenges for biologists and aerodynamicists interested in understanding the design, control and performance of flapping airfoils. New experimental techniques now provide complementary approaches to aerodynamic modeling analyses that are likely to provide new insights into the beauty and complexity of animal flight. Although much work has been carried out to understand the basic features of steady forward flight and the aerodynamic principles that underlie this capability, future work will seek to unravel the mechanisms by which animals maneuver in flight and how the neuromuscular system functions to achieve the impressive array of aerial acrobatics of which both small and large flying animals are capable.

6 | Cell crawling

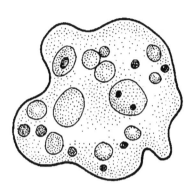

In addition to the ciliary and flagellar propulsion of aquatic unicellular animals discussed in Chapter 4, many other unicellular organisms crawl across the surfaces of substrates. Crawling is achieved by continual reorganization of the organism's cytoskeleton to form surface extensions, or cellular processes, such as lamellipodia or pseudopodia, which are used to adhere the cell to the substrate and to propel it across the surface. These cell-surface extensions have a dense meshwork of actin filaments (homologous to the actin in muscle) located directly beneath the cell membrane (cell 'cortex') that are polymerized and organized in a way that enables free movement of the cell, providing the propulsive force necessary for movement. This involves the protrusion of a cell process, possibly under hydrostatic pressure from within the cell, the formation of a gel-like network of actin filaments which establish mechanical contact with the cell substrate via transmembrane proteins, and the active movement of the cell via a traction force transmitted by this cytoskeletal network. In addition to free-living cells, such as *Amoeba proteus*, many cells within multicellular organisms are also capable of motility. Cell motility is fundamental to many processes of multicellular organisms, such as embryonic development and morphogenesis, tissue repair and remodeling or regeneration, and body defense. In disease, cell motility is also often associated with uncontrolled cell division and growth that leads to metastasis, or the spread of cancerous cells within the body of animals. Finally, cell motility on substrates is probably linked to the same mechanisms that animal cells use to change shape.

The fundamental importance of cytoskeletal organization and how the cytoskeleton mediates changes in cell shape has resulted in an enormous body of research on the genetic control, biophysics and protein biochemistry underlying cytoskeletal organization and function. In addition to providing the mechanism by which single-celled organisms crawl over substrates, the cytoskeleton is key to how many cells sense and interact with their environment and how this affects their functional state. Hence the role of the cytoskeleton in the signal transduction pathways that cells use to communicate with one another and respond to external signals, whether or not they are free living or part of a multicellular organism, has promoted considerable research in this area. Before discussing substrate locomotion of free-living single-celled organisms, such as *Amoeba*, we first need to discuss the basic features of how the cytoskeleton is organized within animal cells. Its use for locomotion will also be discussed in the context of its other roles which include controlling cell organization, cell shape and cell division, as well being an important pathway for cell signaling.

6.1   Organization of the cytoskeleton in animal cells

The cytoskeleton of animal cells consists of three classes of protein filaments: microtubules, actin filaments and intermediate filaments (Fig. 6.1). These are important to a broad range of functions in addition to the cell's structural support and movement. They are briefly reviewed here, but more detailed and extensive discussion of the cell biology of the cytoskeleton is available elsewhere in standard cell and molecular biology texts. Each of these structural elements is highly labile, repeatedly undergoing polymerization and depolymerization over varying timescales. The half-life of microtubules within a cell commonly averages about 10 minutes, whereas the half-life of actin filaments may range from a few minutes to several hours. The ability of these proteins to be turned over and to reorganize is fundamental to the dynamic and changing functions of individual cells.

6.1.1   Microtubules

Microtubules (see Fig. 4.10) are long hollow cylinders constructed of the globular protein **tubulin**, which polymerizes via guanosine triphosphate (GTP) hydrolysis. Tubulin exists as heterodimers of α-tubulin and β-tubulin which polymerize in a repeating and parallel alternating series to form a hollow cylinder. With a diameter of 25 nm, microtubules are more rigid than actin filaments. They are also typically quite long and fairly straight, often extending across the full breadth of a cell. Microtubules within the cell attach at one end to a 'microtubule organizing center', or **centrosome**, which is usually located near the cell nucleus. The centrosome is the primary formation site of the microtubules. Microtubules are polarized polymers, with their 'plus' ends located near the cell surface and their 'minus' ends at the centrosome. This allows microtubules to be used much like a surveyor's scope to locate the cell's center in relation to its outer dimensions, as well as to organize cell organelles (endoplasmic

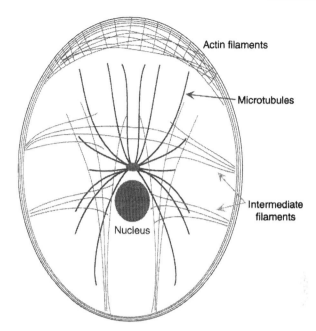

Fig. 6.1 Schematic drawing of a cell showing the general arrangement of actin filaments, which tend to be concentrated within the cortex of the cell beneath the plasma membrane and at the front end of the cell when it moves; microtubules, which form the spindles that emanate from the centriole but also form a network for intracellular transport; and intermediate filaments, which form a layer over the nuclear membrane and provide a general scaffolding for mechanical support of the cell. The microtubules organize into long filaments within cilia and flagella and their relative sliding produces the bending associated with locomotor propulsion of eukaryotic cells (see Chapter 4).

reticulum, Golgi organs, mitochondria) in relation to the cell's metabolic and functional state.

Microtubules also are fundamental to a broad range of intracellular transport functions, particularly when compounds within a cell must be moved long distances, for example in the transport of compounds up or down an axon between its terminus and the nerve cell body. Cellular transport is accomplished by two classes of motor proteins, **kinesin** and **dynein**, which use the microtubules as a transport rail to move their cargo. These motor molecules have similar properties to **myosin-I** found in animal muscles, in that they are capable of coupling ATP hydrolysis to their conformational change for movement and force generation. Whereas kinesins move toward the plus end of the microtubule, dyneins move toward the minus end. Hence transport is mediated in either direction by these two motor proteins.

In addition to intracellular transport, microtubules and their capacity for reorganization are fundamental to how eukaryotic cells divide.

The microtubules form the mitotic 'spindles' (and the centrosome forms the 'spindle pole') of a dividing eukaryotic cell, along which the chromosomes are aligned to ensure an equal distribution of genetic material to the daughter cells. Finally, as we discussed in Chapter 4, microtubules are also fundamental to the organization and function of cilia and flagella. The microtubules of cilia and flagella are organized at their base by a 'basal body', from which they grow. The motor protein dynein is also found within cilia and flagella, and is responsible for the force generation needed for the relative sliding of the '9 + 2' microtubular network (see Chapter 4, section 4.6), which results in bending of the cilia and flagella that produces the drag-based hydrodynamic forces necessary for fluid propulsion.

6.1.2 Actin filaments

The actin filaments found within all eukaryotic cells are homologous to the actin filaments found within animal muscle fibers. As we discussed in Chapter 2, the actin filaments consist of a slender polymer of globular actin monomers organized into a double helix, yielding a filament that is about 5–9 nm in diameter and of varying length. Within different cells, actin filaments can be arranged into a variety of linear arrays (e.g. muscle), two-dimensional networks or three-dimensional gels. Specific cross-linking molecules or 'actin binding proteins' regulate the three-dimensional organization of actin networks and how they contribute to the mechanical properties of a cell. Like microtubules, actin filaments are polarized with a plus and a minus end. When actin filaments are 'decorated' with myosin molecules, the myosins bind at an angle to the actin monomers, giving the actin filament a repeating arrowhead appearance when observed under an electron microscope. The 'barbed' end of a decorated actin filaments corresponds to their plus end.

Although actin filaments are generally dispersed throughout a cell, they are more concentrated and organized beneath the cell's surface. It is this cortical network of actin filaments which underlies cell motility and cell shape change, which will be discussed in detail below. Like microtubules, actin filaments are continually formed and depolymerized within the cytoskeleton of a cell. This contrasts with the much longer half-life of actin filaments found within muscle cells which must form a stable linear lattice to allow repeated contraction based on the cross-bridge interaction between myosin and actin for movement and force (Chapter 2). Actin subunits attach to the barbed (plus) end faster than to the pointed end, and so actin filaments grow in the direction of their barbed ends. The dynamic nature of actin filament nucleation, polymerization (growth) and depolymerization (breakdown) is fundamental to the reorganization and transmission of force that allows cell shape change and movement.

6.1.3   Intermediate filaments

The third class of protein elements which make up the cytoskeleton are the intermediate filaments. Intermediate filaments comprise a heterogeneous family of fibrous proteins including the keratins, which form rope-like filaments with a diameter of about 10 nm. In addition to forming a structural supporting layer beneath the nuclear membrane (known as nuclear laminins), intermediate filaments form a cytoplasmic meshwork of fibers which provide structural support for a cell. Unlike microtubules and actin filaments, intermediate filaments are non-polarized. They consist of rod-like extended monomeric fibrous proteins which dimerize into a coiled-coil structure. These, in turn, form tetramers which assemble into large overlapping arrays that may extend over long distances, for example in the axons of nerve cells in which the neuronal intermediate filaments may be many centimeters long. Intermediate filaments are generally believed to form a tension-resisting skeletal support network for cells, allowing them to resist external mechanical forces. This is especially apparent in epithelial cells which are specialized to produce protective and mechanical structures, such as hair, claws and horns, all of which are constructed from keratin.

6.2   Cell crawling: formation of lamellipodia and pseudopodia for traction and locomotor work

Cells initiate crawling in response to the stimulation of surface receptors located on their membrane. Dynamic surface extensions of motile cells (Fig. 6.2) contain an extensive concentrated network of actin filaments. When these foot-like processes are stubby, such as those produced by crawling amoebae, they are termed **pseudopodia**. Other cells may form sheet-like extensions of their cell surface known as **lamellipodia**, or thin stiff protrusions called **microspikes**. These spike-like projections contain bundles of actin filaments oriented with their plus ends pointing outward. They probably enhance the cell's 'grip' with the substrate, much like the cleats of an athletic shoe on the ground. This is mediated by transmembrane proteins, such as integrin, which mechanically link the actin network inside the cell to the substrate. Substrate adhesion is critical for establishing a traction force that allows for directed progression of the cell (usually in the direction of a chemotactic stimulus). If lamellipodia and microspikes fail to adhere to the underlying substrate, they are quickly swept back over the top of the cell and reabsorbed. The forward moving edge of a motile cell is referred to as its 'leading edge'. It is at the leading edge that transient adhesive contacts with the substrate are formed which enable a cell to pull itself forward.

During extension, the cell process flows locally like a liquid. However, at its leading edge it then forms a stiff gel which resists deformation.

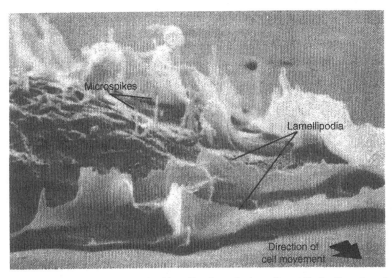

Fig. 6.2 Scanning electron micrograph of lamellipodia and microspikes at the leading edge of a migrating human fibroblast. The direction of cell movement is shown by the arrow. (Courtesy of Julian Heath; copyright permission Garland Publishing Inc.)

The actin-binding protein α-**actinin** cross-links the actin filaments in such a way that an orthogonal network is formed (Fig. 6.3). This provides an efficient space-filling array which maximizes the mechanical integrity of the gel for a given rate of actin filament formation. Another class of actin binding proteins, known as **cofilin** or 'gelsolins', act to dissolve the actin filament network by severing and depolymerizing the actin filaments at their minus (pointed) end. They have been termed gelsolins because of their ability to convert the actin network gel into a solution of freely disrupted shorter actin filaments and actin subunits (Stossel 1993). By undergoing a phase transition from a gel to a solution, the cell is able to pull itself forward after establishing a focal adhesive contact adjacent to where the actin network forms a stiff gel and then to re-extend as the gel is dissolved before forming new adhesive contacts (Fig. 6.4).

6.3 Dynamics of actin nucleation, polymerization and degradation

Locomotor movement of freely mobile cells such as amoeba is achieved through the ongoing formation and degradation of actin filaments. The leading edge of motile cells forms nucleation sites for actin polymerization (Fig. 6.4). Hence, forward movement of the cell is produced by actin polymerization beneath the cell's surface causing the leading edge to advance. The actin

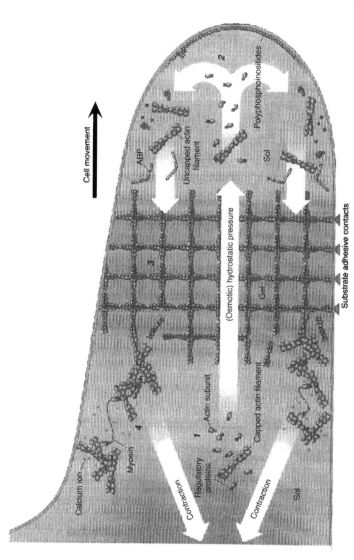

Fig. 6.3 Schematic drawing summarizing the hypothesized mechanism of how the actin network changes from a stiff gel to a fluid (sol), allowing a crawling cell to anchor itself to the substrate via adhesive contacts adjacent to the gel. Myosin motor molecules pull the cell forward by transmitting force and pulling on the actin cytoskeleton. Osmotic pressure (as shown) may serve to extend the leading edge of the cell process and drive dissolved actin network components toward the advancing front of the cell where the actin monomers may initiate the formation of new actin filaments at their barbed ends. Cell surface signaling via polyphosphoinositides directs actin filament nucleation, growth and movement. Other mechanisms underlying cell extension and movement have also been hypothesized. See text for further details. (Adapted from Stossel (1994), with permission from W. H. Freeman.)

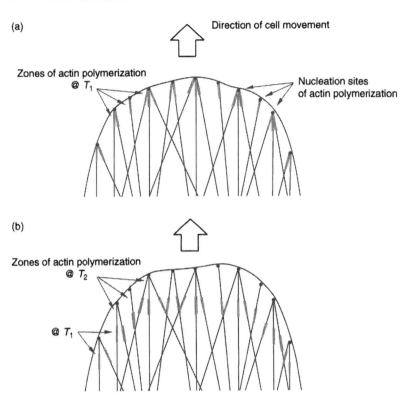

(a)

Direction of cell movement

Zones of actin polymerization
@ T_1

Nucleation sites
of actin polymerization

(b)

Zones of actin polymerization
@ T_2

@ T_1

Fig. 6.4 Schematic drawing showing nucleation sites of actin filament polymerization which predominate beneath the leading edge of a motile cell and the mechanism of cell motility via 'treadmilling' of actin filaments. (a) Actin polymerization which occurs during time T_1 is represented by thicker gray lines along actin filaments. (b) As actin polymerization proceeds (during time T_2), this causes the regions that formed during time T_1 to move in the opposite direction, away of the advancing front of the cell. (Adapted from Theriot and Mitchison (1991).)

filaments are aligned with their plus ends toward the advancing front of the cell and their minus ends at the rear, or cell interior. As noted above, the minus end of the actin filament is relatively inert or slow-growing, whereas the plus ('barbed') end is fast-growing. Growth, or polymerization, of an actin filament is initiated at a site of **nucleation** where actin monomers are induced to form dimers and then more easily trimers, with subsequent polymerization into a longer chain being favored by an increased binding affinity of additional actin monomers. Fragments of disassociated actin filaments also serve as effective nucleation sites. In general, nucleation sites are believed to be common beneath the cortex of a cell's leading edge.

Actin polymerization is sped up by ATP hydrolysis. At the same time, this also enhances depolymerization of the actin filament at its minus end.

This helps to maintain a cytoplasmic pool of actin monomers which can be mobilized for binding to the growing end of the actin filament. The ability of actin filaments to grow at their plus end while being disassembled at their minus end results in what has been described as filament 'treadmilling' (Theriot and Mitchison 1991). This treadmilling behavior of densely organized actin filaments in the cortex of cellular foot-like processes, or pseudopodia, is believed to be a fundamental component of the mechanism that drives cell movement or changes in cell shape (Fig. 6.4 and see below).

In most cells, the actin monomer concentration exceeds the critical concentration necessary for net polymerization. Consequently, in order to control actin filament assembly (in vertebrate fibroblasts roughly 50 per cent of the actin is polymerized into filaments and 50 per cent is in monomer) most of the actin monomers are bound to 'capping' proteins known as **profilin** which regulate their ability to polymerize. Following depolymerization at the trailing edge of the network gel, profilin binds to ADP-bound actin subunits to prevent them from spontaneously assembling into new filaments. The ADP is then exchanged for ATP to 'recharge' the actin monomers, forming an ATP-bound profilin-actin monomer complex. This complex is then recycled to the leading edge of the cell where it can bind to the growing barbed ends of new actin filaments. Exactly how this process is regulated is not yet known. However, regulatory binding proteins are believed to provide a signaling mechanism by which external stimuli are sensed by the cell, initiating actin polymerization and cell movement or shape change. Release or uptake of the regulatory proteins is probably mediated by cell-surface receptors at the leading edge of the cell.

6.4   Cytoskeletal mechanisms of cell movement

Although the association between actin polymerization and cell protrusion is well established, the exact mechanism(s) by which force is produced to extend a cell process, establish a traction force and move the cell forward is not yet clearly understood. One hypothesis is that osmotic swelling of disassembled actin filaments produces the force necessary for protrusion of the cell membrane (Oster and Perelson 1987) prior to the formation of a network gel (Fig. 6.3). Another is that Brownian motion of solutes (principally actin monomers and capping proteins) beneath the cell membrane surface causes expansion of the cell membrane, allowing actin polymerization to advance in a directional fashion (Oster *et al.* 1993). A third is that the dynamic 'treadmilling' behavior of actin filaments which results from polymerization at their front plus end and depolymerization at their rear minus end drives the cell forward (Fig. 6.4) (Theriot and Mitchison 1991). Finally, a fourth and more recent explanation is that myosin motors anchored to the cell surface, or other components of the cytoskeleton, drive the cell process forward by moving along actin filaments

(a)

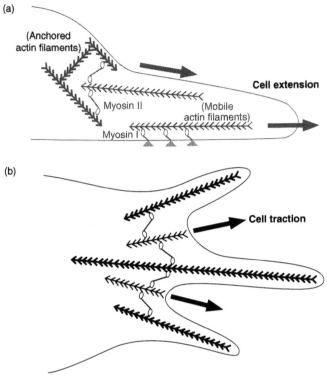

(Anchored actin filaments)

Myosin II

(Mobile actin filaments)

Myosin I

Cell extension

(b)

Cell traction

Fig. 6.5 Schematic drawing showing models of (a) cell extension and (b) cell traction based on myosin I and myosin II molecular motors which drive mobile actin filaments and actin filament networks forward by pulling against anchored sites on the cell. Such models emphasize the importance of myosin for driving cell movements, rather than other forces such as osmotic pressure or Brownian motion at the cell's leading edge. (Adapted from Mitchison and Cramer (1996), with permission from *Cell*.)

which are mechanically reinforced and anchored to local areas of substrate contact (Fig. 6.5) (Sheetz *et al.* 1992; Mitchison and Cramer 1996).

In each of these instances, the underlying mechanism involves nucleation and polymerization of actin filaments beneath the advancing front of the cell. In effect, the formation of actin filaments tracks, and may well assist, forward movement of the cell. It seems necessary that, for net propulsive work to be performed to move the cell forward, the polymerizing actin filaments must be supported by cross-links to other actin filaments and that myosin I and II motors are involved in pulling against local actin networks to pull the cell forward, with one set being anchored to the cell substrate near the leading edge of the cell (Figs. 6.3 and 6.5). This would enable the energy of ATP hydrolysis, coupled to myosin motor work and actin filament growth, to be transformed into active movement of the cell in a given direction.

6.5 Cell-surface receptors mediate sensori-locomotor behavior of unicellular organisms

Unicellular organisms use cell-surface receptors to 'sense' their environment and respond to external stimuli. This involves chemotaxis, in which the cells direct their locomotor movements with respect to chemical concentration gradients, which arise from diffusible chemicals that the cell can sense. When cell-membrane receptors bind the chemical stimuli they trigger localized activation of actin polymerization at sites beneath the cell surface which result in the formation of pseudopodia and lamellipodia, causing the cell to move in the direction of the chemical gradient. For example, amoeba are induced to crawl toward a higher concentration of cyclic AMP, which is a potent chemotactic stimulus of these organisms. The binding of cyclic AMP to cell-surface receptors activates G-proteins, one of the major signal transduction pathways of cells. Activation of the G-proteins results in the release of a protein regulator of the actin monomers, facilitating their nucleation and polymerization at the leading edge of the cell.

6.5.1 The life of a cellular slime-mold amoeba

Dictyostelium discoideum amoebae are free-living single-celled organisms which crawl about feeding on bacteria and yeast. With sufficient food, the cells divide every few hours. However, when their food supply becomes exhausted, the cells stop dividing and merge together to form a small slug-like multicellular structures of size 1–2 cm. These slug-like entities continue to crawl leaving a mucous trail, much like true slugs (Fig. 6.6(a)). Their aggregation is stimulated by the release of cyclic AMP produced by one or more starving amoebae. This creates a chemotactic gradient which directs their movements toward one another. The interaction among individual amoeba cells once they have aggregated and the mechanism by which they coordinate their actin-based locomotor movements is not well understood. During aggregation the amoebae mate, exchanging genetic material. About 2 days after the cells have migrated together for some distance, they begin to differentiate, forming a stalked fruiting body or sporophyte (Fig. 6.6(b)). The fruiting body contains many spores, which can survive for long periods of time even under harsh environmental conditions. When conditions become favorable the spores germinate to produce free-living amoebae which begin the cycle over again.

6.6 Summary

The movements of free-living cells, if not powered in an aquatic environment by cilia or flagella, involve a crawling locomotor mechanism driven by ongoing

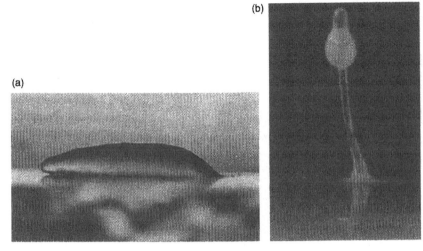

(b)

(a)

Fig. 6.6 Light micrographs of (a) a migrating cellular aggregration, or 'slug', of the cellular slime mold *Dictylostelium* (courtesy of David Francis) and (b) a *Dictylostelium* sporophyte (courtesy of John Bonner). (Copyright permission Garland Publishing Inc.)

actin filament polymerization. Polymerization is powered by ATP hydrolysis, just like skeletal muscle contraction. This occurs principally at nucleation sites beneath the leading edge of the cell. Growth of the actin filaments which may be coupled to myosin motors, as well as osmotic or Brownian forces, exerts a propulsive force that induces the cell to extend itself in the form of lamellipodia, pseudopodia or microspikes. These cellular projections provide traction and a net propulsive force that drives the cell over its substrate. Reaction forces transmitted within the cell are resisted by the cell's cytoskeleton. The cytoskeleton involves microtubules and intermediate filaments which both interact with the actin filaments to support the mechanical forces transmitted via actin filament polymerization. The cytoskeleton serves a broad range of functions other than locomotion and mechanical support. Some of these functions include intracellular transport, control of cell shape and control of cell division.

7 | Jumping, climbing and suspensory locomotion

Many animals jump and several, such as galagos, lemurs, hares, cats, fleas, locusts and grasshoppers, appear to be specialized for jumping. Others, such as frogs, regularly move by jumping. Although jumping does not represent a regular component of locomotion for many animals, the ability to jump is clearly important to terrestrial animals which must cope with obstacles in the environment, pounce to catch their prey or leap to avoid predators. Arboreal animals jump to leap between branch supports for movement in a complex three-dimensional environment. Jumping is also an important component of take-off for flying birds and some ground bats. Additionally, humans and equestrians jump for sport. Although certain types of jumps share similar features to those of hopping, jumping is distinguished from hopping by the fact that it does not require that an animal rebound back into the air after it has landed (one exception being the human triple-jump). Indeed, some anuran species do not land with much grace and balance. This probably reflects the fact that certain frogs often jump into the safety of a pond and toads leap into tufts of grass to hide their position and avoid predators. Presumably in each of these instances the need for a stable balanced landing is of less importance.

Many forms of jumping involve the use of a pair of rapidly extending limbs. For vertebrate quadrupeds and insects this most often involves the use of the hindlimbs. Running jumps, on the other hand, usually involve the use of a single limb. Despite their differences, all jumps commonly reflect two main goals: to jump high or to jump far. As we shall see, simple equations of ballistic motion show that there is a trade-off between jump height and jump length and suggest that there is a particular take-off angle (generally about 45°) for which most jumpers achieve the greatest distance. Jump distance and height are improved by various mechanisms and techniques for jumping. These include longer limbs with larger muscles, catapults which utilize catch mechanisms and elastic energy stored in a spring, and counter-movement of the animal's center-of-mass preceding the jump. We shall review these features after first discussing the basic physics of jumping.

7.1   Jump take-off: generating mechanical power

Jumping requires that animals generate sufficient kinetic energy (KE) to propel themselves into the air. This is accomplished during the take-off phase. Beginning at rest, a jumping animal achieves its maximum kinetic energy at the end of take-off when it leaves the ground (Fig. 7.1). Consequently, to jump as far or as high as possible, an animal must maximize its take-off velocity v_t, giving it a kinetic energy of mv_t^2. This kinetic energy is converted into potential energy (PE) as the animal travels a given distance and height through the air before landing. For vertical jumps the potential energy that an animal attains at the maximum height h of its jump equals the kinetic energy that it achieved at take-off:

$$mgh = 1/2\, mv_V^2 = E \tag{7.1}$$

so that

$$h = v_V^2/2g \qquad h = E/mg \tag{7.2}$$

where v_V is the animal's vertical velocity $m = M$ (body mass) and g is the acceleration due to gravity, which decelerates the animal's vertical motion after it has ceased accelerating during the take-off phase of its jump. Equation (7.2) shows that if an animal can double its take-off velocity, it can achieve a fourfold increase in jump height.

The power for achieving a certain take-off kinetic energy comes mainly from the limb muscles, which must shorten rapidly to maximize their power output. Elastic energy may also be stored in tendons or apodemes, or even possibly in the skeleton itself, and used to amplify the power achieved during take-off beyond that of which the animal's muscles are capable during the acceleration

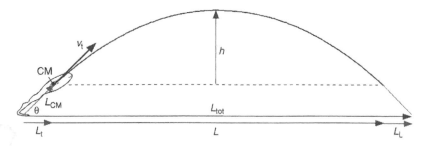

Fig. 7.1 Schematic diagram of a frog jump, with various parameters defined. See text for additional details. (Adapted from Marsh (1994), with permission.)

phase of the jump. This is because the elastic energy can be released from a spring-like element much more rapidly than a muscle can shorten while it generates significant force (see Fig. 2.4(a) in Chapter 2 which shows that as the speed of shortening increases muscle force falls rapidly). In addition to increasing their kinetic energy, animals also have to raise their center-of-mass, increasing their potential energy during the take-off phase of flight. This is often ignored because it is usually a small fraction of the kinetic energy achieved during the take-off. Nevertheless, it is important to bear in mind, particularly for larger animals in which the increase in potential energy during take-off may be a significant component of the total energy increase achieved for a jump.

In general, effects of drag and lift while the animal moves through the air are typically ignored for animals greater than 10 g. Below this size, air resistance can be a considerable hindrance (Bennet-Clark, 1977). The jump height of a flea would be twice as high in a vacuum (zero drag) as it achieves in air. Air resistance consumes 75 per cent of the jump energy that it produces with its limbs. In comparison, a 0.44-g grasshopper loses about 25 per cent of its energy to air resistance. Air resistance is of decreasing importance for larger vertebrate jumpers such as frogs (5 per cent), galagos (3 per cent) and humans (<2 per cent). Lift is not usually considered important during jumping for most animals, unless they have become specialized for gliding (see Chapter 4). In human ski jump competitions, lift is now recognized as a subtle but key factor underlying a jumper's success.

At the end of take-off, the final vertical and horizontal velocities of a jump can be resolved from the animal's take-off velocity and take-off angle θ (Fig. 7.1):

$$v_H = v_t \cos \theta \tag{7.3}$$

$$v_V = v_t \sin \theta. \tag{7.4}$$

The horizontal distance that the animal travels during a jump (returning to its take-off height) is

$$L = v_H t_{air} \tag{7.5}$$

where t_{air} is the time that it spends in the air. Jump height determines the time that the animal spends in the air, which can be derived as

$$t_{air} = \sqrt{(2h/g)}. \tag{7.6}$$

These equations follow from the simple physics of ballistic motion, for which the position(s) of a moving object or projectile at any point in time that is subject to a constant acceleration is defined by $s = vt + at^2/2$, where v is the initial velocity of the object and a is its acceleration). Given that the animal's initial velocity is zero for standing jumps, this simplifies to $s = 1/2 at^2$, from which eqn (7.6) is derived ($a = g$). Using the above equations, the length of a jump (while in the air) can also be calculated as

$$L = (v_t^2 \sin 2\theta)/g \tag{7.7}$$

(recognizing that $2\sin\theta \cos\theta = \sin 2\theta$). This relationship predicts that animals should achieve a maximum jump distance at a take-off angle of 45°. In fact, however, because animals with different limb lengths travel different distances during take-off (L_t) as well as during landing (L_l) (Fig. 7.1), the optimum take-off angle varies for animals with different limb lengths and take-off velocities (Marsh 1994). The above equations govern only the motion of the animal after it has left the ground. They do not account for motions of the animal's center-of-mass during the take-off while it is on the ground. Nor do they account for differences in movement which may occur during landing compared with take-off. Marsh (1994) provides a clear and detailed discussion of these additional components of jump performance. In general, a ballistic analysis yields a reasonably reliable evaluation of jump performance and the various designs which animals have evolved to enhance jumping ability. Nevertheless, it is worth bearing in mind that the distance an animal travels during the take-off (L_t) and landing (L_l) phases of a jump may constitute 20 per cent or more of the total length ($L_{tot} = L + L_t + L_l$) and height of a jump.

The optimum take-off angle for maximum jump length, as well as the angle of force applied to the ground, depend on the magnitude of force exerted by the animal (Fig. 7.2). Because of this, the optimum take-off angle is predicted to be lower for shorter jumps and larger animals. The optimum angle of ground force is always greater than the angle of take-off because a significant component of force is needed to counteract the animal's weight. This is particularly the case for shorter jumps (Fig. 7.2). In general, larger animals can be expected to use lower take-off angles because they exert relatively lower ground forces for their weight (N/mg). For take-off angles in the range 30°–55°, jump length is generally within 90 per cent of more of maximum jump length (Fig. 7.3).

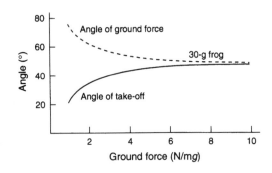

Fig. 7.2 Graph of jump angle and ground force angle which maximize the jump length for a 30-g frog as a function of ground force (Adapted from Marsh (1994), with permission.)

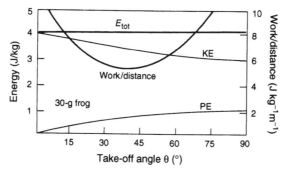

Fig. 7.3 Graph of potential energy (PE), kinetic energy (KE), total energy (E_{tot}) and work/distance as a function of take-off angle for the jump of a 30-g frog. (Adapted from Marsh (1994), with permission.)

Consequently, jump performance is not significantly reduced over a fairly broad range of take-off angles.

Interestingly, the total work (KE + PE) of a jump is almost constant over a broad range of take-off angles and ground force angles. This results from the offsetting changes in kinetic and potential energy required for different take-off angles, in which the more vertical the jump take-off, the greater the PE component and the lower the KE component of work (Fig. 7.3). However, because jump distance varies with take-off angle, the amount of work that a jumping animal's muscles must perform to jump a given distance is strongly affected by take-off angle. Again, the optimum (minimum work/distance) is close to a take-off angle of 35°–40° (with a ground force angle of 45°).

7.2 Scaling of jump performance

In a paper delivered as an Evening Lecture to the Royal Society of London, Nobel laureat A.V. Hill argued, on the basis of geometric similarity and uniform muscle properties, that all muscles can be expected to perform the same amount of work per unit mass (J/kg) and that all animals should jump to the same maximum height and run at the same maximum speed (Hill 1950). This follows from the proportional scaling of length with respect to time ($l \propto t$). Larger animals have longer limbs but move them at slower frequencies ($f \propto t^{-1}$), resulting in the prediction of similar maximum speeds of movement irregardless of body size ($v = l/t \propto m^0$). Similarly, Hill's simple model requires that all muscles have the same maximum speed of shortening v_s. Any change in a muscle's intrinsic shortening velocity (v^*, measured in lengths/s $\propto m^{-1/3}$) is offset by the length of its fibers ($l_f \propto m^{1/3}$):

$$v_s = v^* l_f \propto m^{-1/3} m^{1/3} \propto m^0.$$

Muscle power (work/time) depends on the product of force and velocity ($P = F \times v_s$). Given that muscle force varies with muscle cross-sectional area ($F \propto A \propto m^{2/3}$) and that the maximum speed of muscle shortening is predicted to be constant, this implies

$$P \propto m^{2/3} m^0 \propto m^{2/3}. \tag{7.8}$$

Consequently, as we have seen for other forms of locomotion, mass-specific muscle power P^* (W/kg) is predicted to decrease with increasing size:

$$P^* (P\,m^{-1}) \propto m^{-1/3}. \tag{7.9}$$

This decrease in muscle mass-specific power is offset by the greater take-off times of larger jumpers ($t \propto m^{1/3}$) so that, according to geometric similarity, the total energy ($E^* = P^* t$) derived from the muscles per unit mass of the animal and its muscles should be constant ($E^* \propto m^0$). From eqn (7.2), this again predicts that maximum jump height should be the same for animals of different size assuming that they have the same proportion of muscle mass. Is this the case?

Smaller animals have shorter limbs and because of this the distance and time available for acceleration during take-off are less. If the animal accelerates uniformly (constant ground reaction force) from rest to its take-off velocity v_t over a distance s that is proportional to its leg length, the time required to take-off is

$$t_c = 2s/v_t. \tag{7.10}$$

As a result, smaller animals must achieve greater acceleration a and force to jump a given height, according to

$$a = F/m = v_t^2/2s. \tag{7.11}$$

Larger animals with longer legs are able increase the time that they exert force on the ground accelerating their mass over a greater distance during limb extension. Because of their short legs, small jumpers cannot contract their muscles sufficiently fast to achieve the accelerations necessary to jump as high as larger animals (Bennet-Clark 1977). Consequently, Hill's (1950) simple model for jumping (and running) fails to hold; changes in muscle shortening velocity do not scale independent of size. Nevertheless, Hill's model illustrates several fundamental principles that underlie the scaling of jump performance. In fact, when jump height or distance is compared relative to an animal's length, small jumpers like fleas and locusts are able to achieve dramatic performances compared with large jumpers (Table 7.1). As Bennet-Clark (1977) points out, whereas an antelope need only exert forces 1.6 times its body weight to jump to a height of 2.5 m, a flea jumping 1/25th as high must produce forces 200 times its body weight. Such large forces are required because of the very small size and short take-off time of a flea. Imagine what it must be like to experience an

Table 7.1 Jumping performance of various animals

Animal	Jump height (m)	Take-off distance (m)	Take-off time (ms)	Mean acceleration (g)	Peak power (W/kg)
Antelope (200 kg)	2.5	1.5	430	1.6	115
Human (70 kg)	0.6	0.7	350	1.0	34
Cat (2.5 kg)	1.5	0.32	250	3.2	118
Galago (0.3 kg)	2.25	0.16	100	7.3	410
Cuban tree frog (12.9 g)	0.65	0.108	60	6.1	213
Locust (3 g)	0.45	0.04	26	12.0	330
Flea (0.5 mg)	0.10	0.0005	0.7	208	2750

Data from Bennet-Clark (1977) and other sources.

acceleration of $200g$! Even for astronauts, maximal accelerations during take-off or when re-entering from the Earth's orbit do not exceed $10g$. As we shall see below, fleas make use of a catapult energy storage mechanism to achieve such impressive jumps for their size.

7.2.1   Scaling of jump performance in frogs and toads

Even when compared within a more closely related group of jumpers, jump distance does not remain constant but clearly scales with body size. The maximum jump distances recorded for different species of adult frogs and toads show an allometric pattern (Fig. 7.4), with $L_{tot} \propto M^{0.2}$. Frogs achieve greater jump distances than toads of similar size. This difference can largely be explained by the greater relative hindlimb length of frogs compared with toads (see section 7.2.2). Frogs also jump greater distances at warmer temperatures. This reflects their ectothermic metabolism which allows for increased rates of muscle shortening and more rapid acceleration at warmer body temperatures. As Marsh (1994) notes, the observed interspecific scaling of anuran jumping can be explained largely by the scaling of the hindlimb length of these animals. From the preceding equations, maximum jump length is predicted to scale according to the product of mass-specific power and the distance that the animal's center-of-mass is accelerated during take-off, but raised

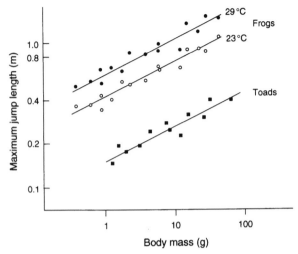

Fig. 7.4 Comparison of jump performance of frogs at two different temperatures and with toads (data from Zug (1978)). (Adapted from Marsh (1994), with permission.)

to the power of 2/3:

$$L_{tot} \propto (P^*L_{CM})^{2/3}. \tag{7.12}$$

Consequently, given that L_{CM} is proportional to hindlimb length, for geometrically similar animals ($l \propto m^{1/3}$) this predicts $L_{tot} \propto (m^{1/3})^{2/3} \propto m^{0.22}$, close to the observed scaling relationship.

7.2.2 Limb length and jump distance

Clearly, limb length is a main factor influencing jump distance. For animals of similar size, species with proportionately longer limbs achieve greater jump distances. This is apparent when comparing the jump performances of frogs and toads (Fig. 7.4). Similarly, Emerson (1985) showed that jumping mammals (marsupial macropods, primate leapers and jumping rodents) have longer limbs than non-jumping mammals of similar size (Fig. 7.5(a)). Both groups exhibit a similar scaling pattern with size ($l \propto m^{1/3}$), but the longer hindlimbs of jumping mammals reflect a clear specialization for increased jump performance. When compared with small jumping mammals (Fig. 7.5(b)), the hindlimb lengths of frogs also scale along the same line. Therefore similar adaptations for increased hindlimb length and jump performance appear to apply broadly across a broad diversity of vertebrate (and invertebrate) species.

To a certain extent, the importance of limb length is also reflected in the jump performance of human athletes. Top human jumpers generally have long legs for their size and weight. However, in human high-jumping the ability to control the motion of the body, arms and legs in relation to the

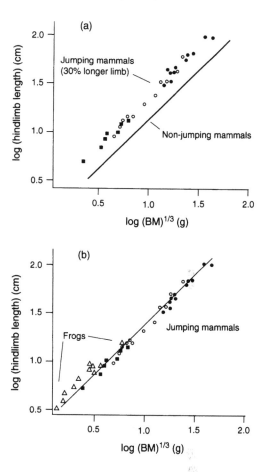

Fig. 7.5 Scaling of hindlimb length in (a) jumping and non-jumping mammals (jumping rodents, solid squares; primate leapers, open circles; kangaroos and wallabies, solid circles) and (b) frogs (open triangles) and jumping mammals (same symbols as in (a)). (Adapted from Emerson (1985), with permission.)

jumper's center-of-mass are equally important for maximizing jump height. In the case of long-jumping, the ability to accelerate and achieve a high kinetic energy which can be translated into the energy of the jump is most critical. The most competitive human long-jumpers are those who achieve high sprint speeds. One of the more impressive examples of this is Carl Lewis, who won *both* the 1984 and 1988 Olympic men's 100-m sprint and long-jump competitions. In both the long jump and the 100-m sprint, Lewis and other top competitors achieve maximum speeds in the range of 10–11 ms^{-1}. This allows them to achieve considerable kinetic energy ($\sim$37 kJ) which can be translated into the horizontal distance of the jump. However, this is not the case for high-jumping, which emphasizes the importance of elastic energy recovery and the power of the leg muscles necessary for translating the body's horizontal kinetic energy into vertical motion. As a result, high-jumpers approach the jump at a relatively slow speed and with a less inclined orientation of their take-off limb (45°–50°) compared with the limb take-off angle of long-jumpers (60°–65°).

In general, specialized jumping vertebrates have massive hindlimbs because of the need to have large muscles which can generate the power and acceleration necessary during take-off. The relative cost of having to accelerate the mass of the limbs themselves does not appear to be very great. Consequently, in addition to having long legs, it is also important to have large limb muscles to jump well.

7.3 Other mechanisms for increasing jump distance

In addition to increased limb length and enlarged muscle mass, jumping animals have evolved other specializations for enhancing jump performance. Most notably, some small jumping insects use a catapult mechanism which allows them to store and release elastic energy in specialized resilin pads or apodemes that greatly increases the power that can be achieved to accelerate their mass during take-off. The ability to store elastic energy while leg muscles develop force relatively slowly and then to release the energy quickly via elastic recoil is an effective mechanism of power amplification. It is likely that elastic energy storage and release is also an important power-amplification mechanism used by larger jumping vertebrates such as frogs and galagos. However, no clear catapult mechanism has yet been observed within vertebrates. In addition, many animals which jump from rest use a counter-movement to stretch their extensor muscles (as well as elastic structures), which enables the muscles to develop force more rapidly and to a greater magnitude, increasing the amount of muscle power that can be developed.

7.3.1 Catapult jumping

Catapult mechanisms have been found in all insects that jump, but have not yet been identified in any vertebrate jumper. Catapults are devices which store energy in a spring, usually by rather slow force development of an extensor muscle, and release the energy rapidly, yielding very high power output to propel a mass (i.e. the animal's body) a considerable distance. Catapults were popular artillery devices used in medieval warfare when laying seige to an enemy's fortress (Gordon 1978). They are also the basis of a child's sling-shot. In the case of locusts and fleas (Bennet-Clark and Lucey 1967; Bennet-Clark 1975), a catapult mechanism is used to store energy in the apodeme of a large extensor muscle and a flexible portion of the animal's cuticle or a rubber-like resilin pad.

Locusts and grasshoppers jump by rapidly extending their long hindlegs. They do this by contracting their large extensor tibiae muscles (Fig. 7.6), which

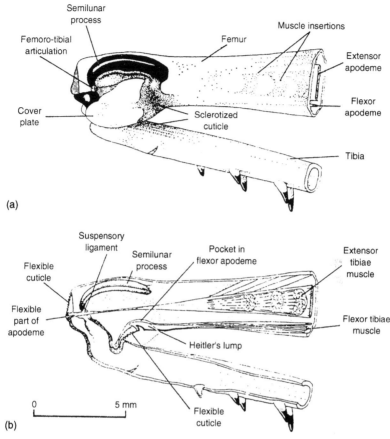

Fig. 7.6 Locust hindlimb showing the knee articulation of the femur and tibia with the large extensor tibiae and much smaller flexor tibiae muscles and their apodemes. (Reproduced from Bennet-Clark (1975), with permission from the Company of Biologists, Ltd.)

are contained inside the stout femur, the large proximal segment of the leg that is comprised of stiff cuticle. The femur articulates with the long and much more slender tibia at the knee joint (as in humans and other land vertebrates). The extensor tibiae is antagonized by a much smaller, but fast-contracting, flexor muscle that also originates from within the femur. Both muscles attach to the tibia via an apodeme, also made of chitin similar to the outer cuticle. On either side of the knee, the femur has a thickening of more flexible cuticle termed the semilunar processes. The tibia is hinged at the distal ends of these processes (Figs. 7.6 and 7.7).

When a locust or grasshopper jumps, the extensor tibiae muscle contracts and develops force, but does not extend the knee. The knee is maintained in a highly flexed position by the antagonist contraction of the flexor tibiae muscle

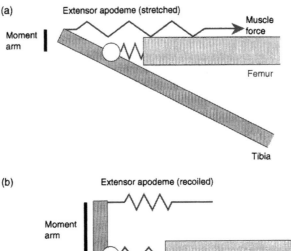

(a)   Extensor apodeme (stretched)

Moment arm

Muscle force

Femur

Tibia

(b)   Extensor apodeme (recoiled)

Moment arm

Semilunar processes (uncompressed)

Fig. 7.7 Schematic model of locust hindlimb showing the catapult and catch mechanisms which underlie the elastic energy release that amplifies power associated with rapid knee extension during a jump. (Adapted from Alexander (1988), with permission.)

which has a much greater moment arm than the extensor muscle allowing it to resist the large force developed by the extensor tibiae. With the knee highly flexed, the apodeme of the extensor muscle passes very close to the axis of rotation of the knee. In addition, Heitler's lump is a bump over which the flexor apodeme passes, forming a catch that helps to keep the knee flexed.

During this time the semilunar processes can be seen to deform as they are compressed, storing elastic energy. In addition, the extensor apodeme is also stretched, storing additional strain energy. These two energy stores are represented by springs in Fig. 7.7. The jump is initiated when the small flexor muscle quickly relaxes, allowing the extensor muscle to shorten. The sudden release of elastic strain energy from the compressed semilunar processes and the stretched extensor apodeme produces a rapid increase in power which results in rapid extension of the knee joint. This occurs much faster (25–30 ms) than the speed of contraction of the extensor muscle, which takes 350 ms or longer to develop maximum tension. In addition, as the knee extends the moment arm of the extensor muscle increases (Fig. 7.7), which allows it to increase the force exerted on the ground. Bennet-Clark (1975) observed that a 1.7-g jumping locust (*Schistocerca*) accelerates from the ground within 30 ms, achieving a take-off velocity of 3.2 m/s and a kinetic energy of 8.7 mJ which is large enough to propel it nearly 1 m. Half this amount of energy is stored in the apodeme and semilunar processes of each hindleg. Fleas accomplish the same task by storing elastic energy in a resilin pad located in the knee joint that is compressed while the extensor muscle rather slowly develops force.

Without the release of stored strain energy from the apodeme and semilunar processes the jump performance of the locust would be greatly diminished. If the locust had to rely solely on the power generated by the rather slow contraction of its extensor tibiae muscle, from eqn (7.10) above a 10-fold longer acceleration time would mean that the locust could only achieve a take-off velocity one-tenth as high, resulting in a jump of only one-hundredth the distance (or 1 cm). Therefore small insects must rely on the rapid release of elastic strain energy via a catapult mechanism to achieve reasonable jump performance.

7.3.2   Power amplification via elastic energy in vertebrate jumping

The release of elastic energy from tendons, muscle aponeuroses and ligaments is also likely to be important to the jumping ability of many vertebrates. Indirect evidence for this comes from the observation that animals such as galagos and some frogs are able to achieve jumps that clearly exceed the power capability of all their hindlimb jumping muscles combined. A 0.3-kg galago which jumps to a height of 2.25 m requires a take-off velocity of 6.64 m/s, which represents energy of 6.62 J (Hall-Craggs 1965). Assuming that the galago achieves a constant acceleration during a 50-ms take-off, this represents an average power of 133 W. Given that the muscles contributing to the animal's jump represent 40 per cent of the animal's body mass, this suggests an average power output of 1108 W/kg muscle. This value is much greater than the peak power output measured for even the very fast glycolytic muscles of frogs (270–371 W/kg) (Lutz and Rome 1994; Marsh and John-Alder 1994). Given that jumping animals cannot develop force instantaneously, the galago's peak power output during a jump must be even greater than 133 W (peak power can be estimated to be roughly twice average power assuming a half-sine pattern of force exerted on the ground (see Chapter 3)). In studies of galago jumping, for which ground reaction forces have been recorded (see section 7.4), Gunther (1985) and Aerts (1998) found that galagos achieved a peak muscle specific power of 1700–1820 W/kg (again assuming the jumping muscles represent 40 per cent of the animal's body mass). Similarly, Peplowski and Marsh (1997) calculated that, in 1.4-m jumps, Cuban tree frogs (*Osteopilus septentrionalis*) achieved an average specific power output of 800 W/kg and a peak power output of up to 1650 W/kg, assuming that all the hindlimb muscles contributed similarly to the power of the jump. These values would be even higher if only a fraction of the total hindlimb jumping muscles were active.

The exceedingly high power outputs that these jumps represent are well beyond the capacity of the animal's muscles, even if all the jumping muscles contracted optimally to maximize their power output. Peplowski and Marsh (1997) measured an *in vitro* peak power output of 230 W/kg by the sartorius muscle of Cuban tree frogs, which probably has contractile properties that are similar to those of the jumping muscles of the hindlimb. Even making the conservative assumption that all of the muscles contracted optimally to contribute power for the jump, this is a factor of 7 less than the peak power indicated by the ballistics of the animal's jump. This suggests that power is amplified by means of elastic energy storage and release in spring elements of the limb. Aerts (1998) estimates a power amplification of 15-fold in the jump of a galago. It must be the case that, as limb extensor muscles develop force with the limb in a flexed position, energy is stored in muscle apoeneuroses, tendons and ligaments. This energy is then rapidly released near the end of take-off as the limb extends, similar to the release of energy that powers the jump of locusts and fleas. However, the exact sites and amounts of elastic strain energy recovery available for powering vertebrate jumpers are still not well identified. Aerts (1998) argues that, in the galago, the knee extensor muscles are probably an important source, along with the ankle extensors and possibly foot tendons and ligaments. Similarly, in frogs and toads, the cruralis (knee extensor) and plantaris longus (ankle extensor) are the two largest muscles of the hindleg. Both are pinnate and both attach to the frog's skeleton via substantial aponeuroses and tendons. It seems likely that energy stored in these collagen-based tissues also functions to amplify the power originally generated by muscle contraction.

Although no catch mechanisms have been found in vertebrate jumpers, it seems likely that changes in muscle mechanical advantage underlie the means by which the limb can be held in a flexed position, allowing the large extensor muscles to develop force under near-isometric conditions and store energy in their elastic structures. Indeed, in frogs jumping from rest, the hindlimb extensor muscles are activated well before there is any sign of movement by the animal's limb. This is a clear indication that the muscles are developing force isometrically and storing energy in associated elastic elements. As in the case of the release of the catch in the jump of a locust when the flexor muscle relaxes, the antagonistic activity of two-joint muscles that span the knee and ankle joints of vertebrate jumpers probably controls the timing of joint and limb extension. If the large extensor jumping muscles operate at a poor mechanical advantage early in the jump, so that they develop force under isometric conditions, a shift in joint moment balance may allow these muscles to begin to extend the limb joints. If joint extension is accompanied with an increase in the moment arms of the extensor muscles (similar to the locust knee extensor), this would allow sudden shortening of the extensors and elastic recoil of the spring elements, providing rapid extension and amplification of power. Future work is needed to identify such mechanisms which are almost certain to exist.

7.3.3 Counter-movement jumping

Although frogs and toads jump from a stationary position, most jumping mammals and birds use a counter-movement during standing jumps to increase their performance. A counter-movement represents an initial flexion of the limb, which lowers the body's center-of-mass. During the counter-movement the force exerted on the ground briefly falls below the animal's body weight (Fig. 7.8(a)). This is immediately followed by rapid extension of the limb to propel the animal into the air. By performing a counter-movement, the jumping muscles are forcibly stretched while they are being activated. This allows the muscles to develop force more rapidly and to a greater magnitude (see Chapter 2, Fig. 2.4(a)). Because of this, the muscles can produce greater power when they subsequently shorten to extend the limb. As we have seen, this pattern of active muscle stretch prior to shortening is a common behavior of many muscles involved in a wide range of locomotor activities. It occurs in the leg muscles of hopping wallabies and kangaroos, the leg muscles of running turkeys, dogs, horses and humans (Chapter 3), and even in the flight muscles of birds (Chapter 5). Humans making a counter-movement in squat jumps achieve greater heights than if they jump from a stationary squat position. In addition to increasing the power output of the muscles, the counter-movement also enables more elastic energy to be stored and recovered during the jump.

Counter-movement jumps from rest use mechanisms similar to those involved in running jumps. When an animal jumps following a running start, it can store elastic energy and forcibly stretch its muscles when the foot is planted before springing into the air. In addition to muscle stretch and elastic energy storage, running jumps also utilize the conservation of an animal's horizontal kinetic energy. The combination of kinetic energy transfer, muscle stretch and elastic energy recoil are all important components for achieving greater power in running jumps. Various combinations of these are critical

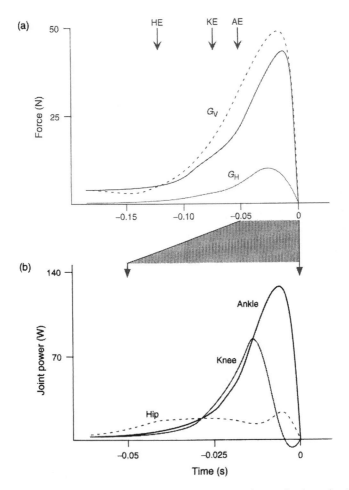

Fig. 7.8 (a) Ground reaction forces developed over the course of a standing jump (vertical and horizontal components, G_V and G_H). The solid curve shows a case in which no counter-movement occurs. The broken curve shows the pattern of force developed when a counter-movement is made (G_V only; G_H is similar in both cases). These patterns of force development are characteristic for a broad variety of vertebrate jumpers. The characteristic timings of the onset of hip extension (HE), knee extension (KE) and ankle extension (AE) are indicated by arrows. (b) The pattern and timing of hindlimb joint power (moment multiplied by angular velocity) during the jump of a galago (Aerts, 1998) show that power is developed in a proximal to distal sequence, with the peaks in knee and then ankle power matching the timing of peak propulsive ground force. Note that the timescale in (b) is expanded relative to that in (a).

to the success of human long- and high-jumpers. Whereas horizontal kinetic energy is probably most critical to long-jumpers, the ability to amplify muscle power through active stretch and elastic energy recoil may be most important in high-jumping.

7.4   Other morphological adaptations for jumping

In addition to enlarged hindlimbs and catapult elastic energy mechanisms, jumping animals show a range of other morphological specializations for jumping (Emerson 1985). The propulsive thrust of the hindlimbs must act through the center-of-mass of the animal's body to avoid unwanted torques which might cause the animal to pitch backward or forward during the jump. Many mammalian jumpers minimize the possibility for pitching moments by their center-of-mass being more posterior so that it is aligned more closely with the propulsive thrust of the hindlimbs. This shift is accomplished in a variety of ways: by reducing the size of the forelimbs and increasing the size of the hindlimbs, by shortening the trunk, and by having a longer and heavier tail. Bipedal rodents, such as jerboas and kangaroo rats, also have fused cervical vertebrae with an enlarged vertebral spinous process to which an enlarged dorsal flexor muscle of the head attaches. This is believed to reduce head bobbing during a jump. Frogs have a shortened presacral vertebral column but, in contrast with mammalian jumpers, lack a tail. To align their trunk parallel with the vector of ground force, frogs rotate their anterior trunk via a presacral hinge that is controlled by muscles that attach between the ilium and sacrum (Emerson 1985). Alignment of the trunk is also facilitated by hip extension, which precedes the onset of knee and ankle extension. This pattern of joint extension is common for a variety of jumping vertebrates (Fig. 7.8(b)).

7.5   Ground forces and joint power underlying vertebrate jumping

When an animal jumps from the ground it exerts a force which rapidly exceeds its body weight (Fig. 7.8(a)). Initially the rise in force is slow, but progressively increases at a faster rate (increased slope). Ground force typically peaks late in the jump just prior to take-off. This pattern of force development is exhibited by a broad variety of jumping animals. Unlike the constant take-off acceleration assumed in simple ballistics equations, these patterns show that the acceleration of the animal's mass is not constant, but peaks late in the jump. This is why the peak power output achieved by many jumpers is much greater than their average power output, demonstrating the importance of elastic energy storage as a means for achieving the high peak power outputs which often greatly exceed the capacity of the animal's muscles. When an animal makes a counter-movement, the force briefly falls below its body weight (broken curve in Fig. 7.8(a)) but this allows it to achieve a greater maximal force output.

The pattern of ground force is also reflected in the pattern of joint power developed within the hindlimb of vertebrate jumpers (Fig. 7.8(b)). Characteristically, the hindlimb joints extend in a proximal to distal sequence: hip extension occurs early in take-off, followed by knee extension and then ankle extension near the end of the jump. This pattern allows muscle power to be transmitted within the limb to the ground. It also allows the large knee and ankle extensor muscles to develop considerable force prior to shortening. When the joint extends, muscle force (or the joint moment) falls rapidly. This coincides with the timing of maximum ground force and its sharp decline as the animal leaves the ground. The late peak in knee and ankle joint power is also consistent with the possibility that energy stored in the elastic elements which these extensor muscles stretch is rapidly released to enable the animal to achieve power outputs which far exceed the capacity of the muscles alone.

7.6 Climbing

The ability to climb is important to many animals, particularly those that are arboreal or must move over irregular and steeply sloping surfaces. The latter is often the case for smaller animals. Because tree branches are limited and variable in width, discontinuous and variable in their orientation, many animals have evolved specializations for moving effectively over such supports. One advantage of being small is that smaller animals can move more readily along branches of a certain diameter as if they were flat surfaces. However, in order to move up vertical surfaces an animal of any size must be capable of generating a vertical reaction force between itself and its substrate that is equal to or greater than its weight. Such a force must act tangential to the support surface. There are two basic means by which an animal can accomplish this: either it can achieve an interlocking surface with the substrate to generate a new non-vertical contact surface between itself and its support, or it can develop an adhesive or suction force between its body and the contact surface (Cartmill 1985). The first mechanism is used by animals that have evolved claws to cling to support surfaces when they climb and the second by animals that have evolved specialized foot pads for gripping the surface. A friction grip can involve both mechanisms.

The ability to balance over a support requires that an animal maintain its center-of-mass in line with its support. This is more difficult for large climbing animals and for smaller supports because the potential for developing a toppling moment is greater. The toppling moment Wd_h equals the product of the animal's body weight W and the horizontal distance d_h from its center-of-mass to the vertical axis of the support. To avoid falling when a toppling moment develops an animal must be able to exert a counteracting torque by achieving sufficient grip on the support surface. Climbing animals reduce their risk of developing large toppling moments by having evolved shorter limbs (for a given angle of pitch to either side of the support axis, this reduces d_h), moving with more crouched postures (Schmitt 1999) to bring their center-of-mass closer to the support axis (also effectively reducing d_h) or being small. However, small size increases the difficulty of clinging to larger-diameter vertical supports when climbing. A problem of large size is that the strength of branch supports becomes limiting. Large animals have difficulty climbing on slender terminal branches, where food resources are often found. This can be mitigated to a certain extent by distributing the support of body weight over multiple supports. Nevertheless, the largest animals that habitually forage in trees are orangutans which do not exceed 90 kg in weight. Other arboreal specialists climb by hanging upside down which ensures that their center-of-mass is suspended in line with their support (see below). Many climbing animals, particularly primates and carnivorans, also have prehensile tails that are capable of gripping the primary or an adjacent support to better resist toppling moments.

Without an interlocking, adhesive or suction grip, in order not to slip when climbing an inclined surface, a static frictional grip F_{fr} must be achieved which supports an animal's body weight:

$$F_{fr} = W \eta \cos \alpha \qquad (7.13)$$

where F_{fr} is the frictional force, η is represents the coefficient of friction between the animal's foot and the support and α is the angle of inclination of the support (Fig. 7.9). This condition is met when the tangential force (T) due to body weight is less than the frictional force:

$$W \sin \alpha < W \eta \cos \alpha \qquad (7.14)$$

or $\tan \alpha < \eta$. This is impossible for a vertical surface ($\alpha = 90°$), and the animal must slip. With a coefficient of friction of 0.36, the value of wood on leather, an animal could climb up a branch with a slope of about 20° by friction alone before slipping (Cartmill 1985).

Similarly, when grasping a circular support with the digits of the hand or foot, a clawless animal exerts an adductor force F_{add} which produces tangential ($F_{add} \sin \beta$) and normal ($F_{add} \cos \beta$) components of force, where $\beta = (180° - \theta)/2$ and θ is the angle subtended by the two points of grip (Fig. 7.10) (Cartmill 1985). When $\theta = 180°$, F_{add} is normal to the support surface; when $\theta = 0°$, F_{add} is completely tangential and no frictional grip is possible. Similar to establishing a frictional grip on a sloping surface, when $\tan \beta = \eta$, the animal's grip will fail. In practice, animals must achieve grip angles θ greater than the theoretical minimum because some fraction of their weight will also act tangentially to the support surface, and this must also be effectively supported. In addition, when one surface is smooth and the other curved, with surface properties similar to skin, the static coefficient of friction can decrease with increasing normal force (Cartmill 1985). Consequently, this also requires a greater angle for effective grip.

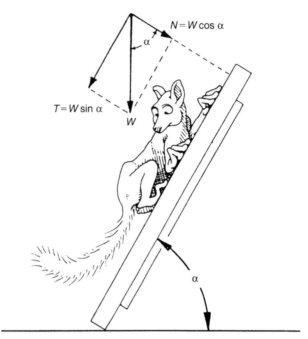

Fig. 7.9 The frictional resistance associated with an animal's weight on an inclined slope depends on the angle of the slope in relation to its static coefficient of friction (see text for additional details). (Reprinted by permission of Wiley-Liss, Inc., a subsidiary of John Wiley & Sons, Inc.)

Rather than relying solely on friction, most arboreal specialists have evolved an array of morphological adaptations that allow them to achieve even greater grip forces. The most straightforward of these is to cling by developing an interlocking grip. Typically, this involves the use of claws which penetrate into the surface of the substrate. By doing this, the claws create a new contact surface that is more nearly perpendicular to the gripping adductor force of their digits. This greatly increases the effective gripping angle θ, which allows them to climb vertical or even overhanging surfaces. Many animals have highly recurved claws with sharp tips, compared with their terrestrial relatives which have blunter and more gently curved claws (Cartmill 1985). This further increases their effective grip angle.

In addition to interlocking grips using claws, many reptiles and amphibians use capillary adhesion to adhere to highly inclined substrates. Specialized pads on the base of the digits of these animals secrete mucous which provides a fluid film that supports the animal's weight. This is achieved by means of the surface tension developed between the fluid and the contact surface. Flattening and expansion of these pads increases the area of contact and the surface tension that can be developed.

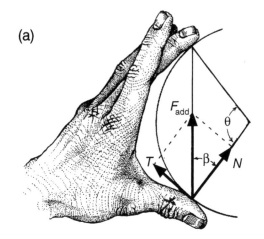

(a)

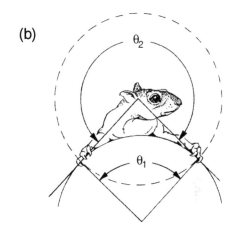

(b)

Fig. 7.10 a) A frictional grip of a circular support depends on the angle subtended by the two points of grip and the coefficient of friction between the animal's grip and the substrate. With $\theta = 180°$, F_{add} is normal to the support surface giving maximal grip. b) Animals with claws can subtend a greater effective angle (θ_2) than that achieved based on friction alone (θ_1) which is determined solely by their points of contact with a branch, as in (A) above (Reprinted by permission of Wiley-Liss, Inc., a subsidiary of John Wiley & Sons, Inc.)

Finally, some reptiles (most notably geckos) and insects rely on dry adhesion to achieve remarkable gripping forces for climbing. The feet of geckos have adhesive pads which consist of small 0.1-mm projections, or setae. Individual setae have hundreds of finer hair-like projections which terminate on 0.2–0.5-μm spatula-like structures (Russell 1975). The foot of a Tokay gecko has about 5000 setae/mm^2. Dry adhesion involves the development of intermolecular forces between the two surfaces. To be effective, such forces must operate at extremely close range (<0.5 nm). These forces arise from the coupling of electron motions in the electron orbitals surrounding adjacent molecules. Recent direct force measurements obtained from isolated individual setae (Autumn *et al.* 2000) provide strong evidence that intermolecular van der Waals forces are the basis for the extraordinary adhesive capability of geckos. In contrast with friction, which typically diminishes when sliding movement occurs because the dynamic coefficient of friction is less than the static coefficient, the resistive adhesive force produced by gecko setae actually increases when the spatular tips begin to slide across a substrate. In addition, complex movements of the adhesive pad underlie the ability of geckos to establish an effective adhesive bond and then to release it, allowing them to rapidly move up smooth vertical surfaces. As the foot is placed on the substrate, the adhesive toe pad is unfurled along its length to contact the surface. As the gecko moves over its foot support the adhesive toe pad peels up, away from the substrate surface, enabling the gecko to break its adhesive bond. The peeling action of the toe changes the orientation of the gecko setae, which presumably disrupts the van der Waals forces established between the spatular tips and the substrate surface.

7.7 Suspensory locomotion at larger size

Various primate species utilize suspensory locomotion to swing through the forest canopy. This is most specialized in gibbons, siamangs and spider monkeys. In addition to continuous-contact swinging, these animals are also capable of richochetal brachiation, which involves aerial flight phases interspersed between successive suspensory support phases. Suspensory locomotion in primates has most generally been modeled as simple pendular locomotion because of the obvious analogy with a freely swinging pendulum (Preuschoft and Demes 1984). Akin to the inverted pendular model of terrestrial walking gaits, pendular motion results from the out-of-phase exchange of potential and kinetic energy of the animal's body as it swings from one overhead contact to another. However, this model greatly oversimplifies the actual suspensory motion of gibbons and other brachiators. It does not account for the richochetal aerial phases of flight, nor does it allow for much of a speed range (Swartz 1989). A simple pendular model requires a narrow range of swinging frequency, which depends on the length of the pendulum in relation to gravity. Since the length of the forelimb, from which the animal is suspended, is relatively constant, the frequency of oscillation and speed of movement are expected to be rather uniform.

In fact, brachiating animals move over a fairly wide range of speeds and must regularly adjust the relative spacing of their overhead limb supports (Preuschoft and Demes 1984). Also, they do not use symmetric motions of their upper limbs and limb joints throughout the suspensory support phase. Recent experiments based on measurements of the reaction forces that gibbons exert on overhead supports (Chang *et al.* 1997) show that gibbons do not simply rely on potential and kinetic energy exchange during pendular support, but they can also throw themselves into the air to achieve greater velocities and stride lengths. This reflects the fact that the arm is not used as a simple cord, but can actively control the momentum and energy of the animal's moving body while it is in contact with an overhead support. Another interesting difference between brachiation and terrestrial locomotion is the difference in the relative timing of horizontal deceleration and acceleration of the animal's body. Whereas terrestrial animals decelerate through the first half of limb support and re-accelerate during the second half (Chapter 3), brachiating animals accelerate during the first half of the swing and decelerate during the second half. This is consistent with the underlying difference in the pendular mechanics of their motion (suspended versus 'inverted'). Recent modeling (Bertram *et al.* 1999) indicates that both continuous pendular contact and richochetal support gaits can provide comparable high mechanical efficiencies. However, the only available oxygen consumption data to date indicate that the cost of locomotion is greater in spider monkeys when they brachiate than when they walk quadrupedally (Parson and Taylor 1977).

7.8 Summary

This chapter has considered more specialized forms of locomotion. Jumping is used broadly by many animals as part of their locomotor repertoire. Its diverse use and parallel evolution in both large and small animals attests to its selective value for avoiding predation, catching prey and moving over obstacles in the natural landscape. Simple ballistic equations of motion predict well the jumping mechanics of animals. However, they neglect important aspects of hindlimb extension during take-off and landing. Indeed, almost all of the attention has been directed toward the take-off and how this affects jump performance. Key morphological adaptations which can be linked to selection favoring increased jump performance are easily observed. Most notably these include enlarged long hindlimbs with more powerful muscles. At small size, jump performance becomes limited by the time required to take off. Because the muscles of small animals cannot contract fast enough, several species have evolved energy-saving catapult mechanisms. This allows their extensor muscles to contract more slowly and to store elastic energy in spring elements of their limbs. The sudden release of elastic strain energy late in the jump take-off greatly amplifies the power and performance of the jump. Strain energy release to amplify jump power is also probably used in jumping frogs and galagos. Although different-sized jumpers cannot achieve similar jump performance, as simple isometric models of muscle contraction suggest, small jumping insects can achieve very impressive jumps when normalized to their body length.

Climbing and suspensory locomotion represent two other specialized modes of locomotion that are well suited to an arboreal environment. A number of morphological specializations, including coarse pads, claws and adhesive pads, enhance the frictional or adhesive contact that a climbing animal can achieve with the surface of its substrate. Perhaps most impressive are the setae of gecko toe adhesive pads, which appear to interact with the substrate surface by means of van der Waals forces, allowing them to scale vertical walls and run upside down on ceilings with great ease. Larger climbing animals, such as primates, have the dual problem of limited branch strength for support and the increased risk of toppling moments. More crouched postures, grasping hands and feet, and prehensile tails are key adaptations which improve balance and enable these animals to counteract toppling moments. Suspensory locomotion at larger size is observed in the richochetal brachiation of gibbons and spider monkeys. Although movement of the body as it swings by the arm appears pendular-like, in fact the motion is not well modeled by a simple pendulum. Instead, these animals can impart energy into the motion of their body during suspensory support to adjust their speed and spacing of overhead supports. Future work in these and other areas will undoubtedly yield fascinating new insights into the varied ways that animals have evolved to move about.

8 | Metabolic pathways for fueling locomotion

Most of the energy that animals use to fuel their metabolic demands is produced in the form of adenosine triphosphate (ATP). Hence, ATP is the ultimate energy source for muscle. Whereas ATP is produced through the catabolic breakdown of nutritional energy reserves (lipid, carbohydrate and less often, protein), the interconversion of metabolic substrates and the storage of energy in the body is achieved through a related set of anabolic, or synthetic, pathways. The metabolic enzymes which catalyze and regulate these reactions in the cells of animals generally have been highly conserved over the course of evolution. Consequently, the pathways by which ATP is produced to provide the energy substrate for motor activity is broadly shared by most animals. This commonality allows us to discuss and compare the metabolic strategies which underlie locomotor activities of many diverse animals.

General principles will emerge for how animals sustain their activity. On the other hand, differences in metabolic capacity for producing ATP are reflected in the different types of fibers which make up an animal's muscles and hence affect its locomotor activity. Balancing energy demands in relation to its reserves and energy input is critical to the success of any organism. Consequently, this need has resulted in the evolution of locomotor mechanisms which conserve an animal's energy expenditure in order to improve locomotor economy. Finally, an animal's ability to fuel its locomotor needs by aerobic metabolism will necessarily link the design and performance of its cardiorespiratory systems for delivering oxygen to the mitochondria within its muscles (Weibel *et al.* 1981). These functional links in structural design among interacting physiological systems reflect the integrated nature of how all organisms work.

8.1 ATP: currency for converting chemical energy into mechanical work

ATP (Fig. 8.1) is by far the most common energy-liberating compound which organisms use to convert chemical energy into mechanical work. Energy is

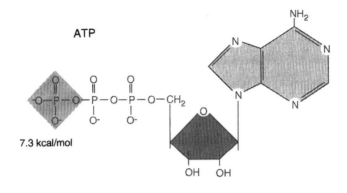

ATP

7.3 kcal/mol

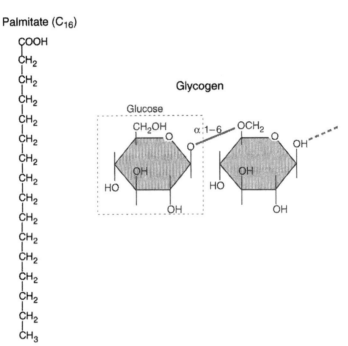

Palmitate (C$_{16}$)

Glycogen

Glucose

Fig. 8.1 Chemical structure of ATP showing the terminal high-energy phosphate group which is split via ATP hydrolysis; glucose, which is stored as glycogen in animal cells by means of an α 1–6 hydroxyl bond between adjacent glucose monomers; and palmitate, a 16-carbon fatty acid.

released by the splitting (via hydrolysis) of the terminal phosphate group (Pi) according to the following reaction:

$$\text{ATP} \rightarrow \text{ADP} + \text{Pi} + 7.3\,\text{kcal/mol ATP}. \tag{8.1}$$

This energy is then available for generating force or producing movement via muscles, cytoskeletal elements, flagella or cilia. It may also be used for active membrane transport or, through various anabolic pathways, to build compounds and tissue structures needed by the organism. Enzymes that catalyze the splitting of ATP are referred to as 'ATPases'. Consequently, in the case of muscle, myosin serves as a myosin-ATPase catalyzing the above reaction. As discussed in Chapter 2, fast-twitch fibers have high myosin-ATPase rates, whereas slow-twitch fibers have slow rates. This difference constitutes a key feature which distinguishes them as histochemical fiber types. In order to fuel the conversion of chemical energy to mechanical work by muscle and other motility-based systems, a readily available supply of ATP must be generated. Two general pathways exist for producing ATP molecules: **aerobic** pathways, which require the use of oxygen, and **anaerobic** pathways, which do not. For most motile organisms, sustained activity is ultimately achieved through aerobic metabolic pathways. Anaerobic metabolism provides a rapid supply of ATP to initiate activity or to facilitate high levels of activity, but the end-products of anaerobic metabolism inevitably lead to metabolic **acidosis** (reduced pH of the cells and internal fluids) and fatigue. Consequently, anaerobic metabolism is considered to be 'non-sustainable'. Ultimately, the end-products of anaerobic metabolism must be reconverted or broken down by aerobic pathways during a 'recovery period' during which the organism is largely inactive. Both sets of metabolic pathways for producing ATP are important to organisms and the manner in which they regulate their activity.

8.2 Aerobic metabolism: oxygen consumption

The net energy and pH balance of most organisms ultimately depends on the consumption of oxygen to fuel ATP production. Consequently, measurements of an animal's oxygen consumption represent one of the fundamental physiological methods for studying the locomotor activity, time course of energy demand and limits of energy supply in various animals. Methods for measuring an animal's oxygen consumption are discussed in section 8.5. The principal sources of energy used by most organisms to fuel their activity are carbohydrates (sugars and starches) and fats. On a per gram basis, the energy yield of carbohydrate (4.2 kcal/g) is less than half that of fat (9.4 kcal/g). However, when expressed in terms of the amount of oxygen required to oxidize each fuel, the amount of energy produced is fairly similar (carbohydrate, 5.0 kcal/1 O_2; fat, 4.7 kcal/1 O_2). Because fat yields more energy per unit weight, it provides the most efficient storage for long-term energy needs. Hence, fat is commonly stored and used for such activities as migration and hibernation. Its disadvantage, compared with carbohydrate, is that fat requires a longer time to be mobilized from adipose storage sites within the body. It can be mobilized more quickly when stored as lipid droplets within the cell itself. Birds and other animals have relatively high levels of lipid within their muscle cells, reflecting their ability to burn fat more readily than the muscles of other animals, such as humans. In general, carbohydrate is utilized first by most organisms to initiate the production of ATP (following the initial use of intracellular high-energy phosphate stores, such as creatine phosphate). Animals which specialize in the economical use of fat for long-term energy supply, such as migratory birds, have evolved specialized proteins (adipose binding proteins) to facilitate the breakdown and transport of fatty acids from adipose storage sites for fatty acid oxidation in the liver.

Carbohydrate is most commonly stored in the form of glycogen within the cell. Glycogen consists of a polymer chain of glucose units (Fig. 8.1). Individual glucose molecules are cleaved off from the polymer chain and ultimately oxidized to form carbon dioxide and water. In the process 38 ATP molecules are produced:

$$C_6H_{12}O_6 + 6O_2 \rightarrow 6CO_2 + 6H_2O + 38ATP. \tag{8.2}$$

The above equation represents the summary balance of two metabolic sequences: the first sequence is **anaerobic glycolysis** and the second is **oxidative phosphorylation**, which occurs within the mitochondria, the ultimate site of oxidation within the cell:

6C-glucose → anerobic glycolysis → (2) 3C-pyruvate

→ oxidative phosphorylation → ATP.

Catabolic products of fat breakdown enter the glycolytic pathway 'downstream' from glucose, but follow the same oxidative pathway within the mitochondria for ATP production. These are briefly summarized below. More detailed presentation of the individual steps of cellular metabolism are contained within biochemistry textbooks.

8.3 Glycolysis: anaerobic metabolism

Glycolysis occurs within the cytosol of the cell and comprises the first sequence of steps in the metabolic breakdown of glucose (Fig. 8.2). Typically, glucose is stored within cells in the form of glycogen, which comprises a long-chain polymer of glucose units. (When derived from glycogen, inorganic phosphate can be combined directly to form glucose-6P without ATP consumption.) ATP production by glycolysis does not require oxygen and hence represents the anaerobic component of cellular metabolism. Through a series of steps, one (6-carbon) glucose molecule is converted into two (3-carbon) pyruvate molecules. If sufficient oxygen is available within the cell, the pyruvate diffuses into the mitochondria and is converted to acetyl coenzyme A (acetyl CoA), which enters the **citric acid cycle** (also often referred to as the **Krebs cycle** or the **tricarboxylic acid cycle**). However, if oxygen is insufficient or

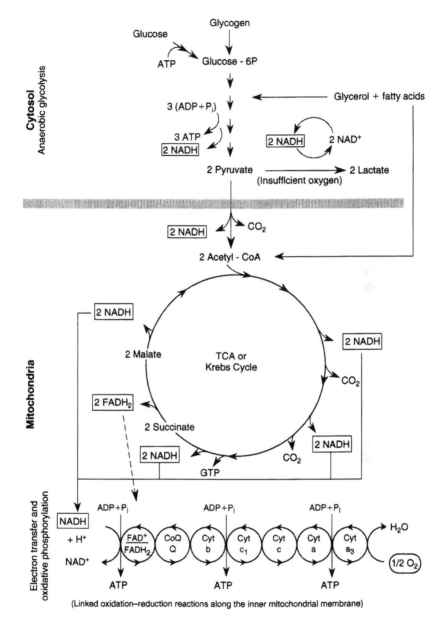

Fig. 8.2 Diagram of the principal pathways for carbohydrate (glycogen or glucose) and fat metabolism within the cells of animals. Anaerobic breakdown of glucose (and glycerol) occurs via glycolysis within the cytosol of the cell. If pyruvate cannot be further oxidized, it is converted to lactate. When sufficient oxygen is available, oxidation of pyruvate (the end-product of glycolysis) and fatty acids occurs through the tricarboxylic acid cycle and oxidative phosphorylation via electron transport within the cytochrome chain of mitochondria. See text for further details.

absent, the pyruvate is converted into lactate (i.e. lactic acid). In the process, one ATP is produced per lactate molecule. Consequently, the net chemical equation for anaerobic glycolysis is

$$C_6H_{12}O_6 \rightarrow 2C_3H_6O_3 + 2ATP. \tag{8.3}$$

(If glycogen is broken down to form pyruvate an additional ATP molecule is produced, providing a net gain of three ATP molecules). Whereas the production of ATP by anaerobic glycolysis is fast, its energy yield is poor compared with aerobic metabolism. In addition, the two lactates ($C_3H_6O_3$) produced must be chemically buffered or else they will depress the pH within the cell (leading to a state of cellular, and ultimately systemic, 'metabolic acidosis'). In multicellular organisms, the lactate initially diffuses out of the cell into the bloodstream or internal fluid compartments, where it can be more effectively buffered. However, excess production of lactate resulting from prolonged anaerobic metabolism ultimately leads to acidosis, fatigue and a rapid decline in activity. Although the production of ATP through the formation of lactate does not have a high energy yield, the remaining energy contained within lactate is *not* lost to the animal. The lactic acid can either be converted from a less toxic form back into glucose for storage or, when sufficient oxygen is available (as during recovery metabolism), it can be reconverted to pyruvate, and the pyruvate can be oxidized to produce ATP via subsequent aerobic pathways (see below). The net cost to an animal resulting from the short-term conversion of glucose to lactate is not that great. The main penalty is the effect that this can have on its pH and probably ensuing inactivity that must follow during metabolic recovery.

8.4 Mitochondria: citric acid cycle and cytochrome oxidative phosphorylation

When sufficient oxygen is available, the pyruvate produced by glycolysis diffuses inside the mitochondria and is transformed into acetyl CoA in the mitochondrial matrix (Fig. 8.3). In the process, carbon dioxide is released and an NADH molecule is formed (two NADH molecules for each pair of pyruvates formed from one glucose molecule). The acetyl CoA then enters the citric acid or Krebs cycle (Fig. 8.2), which also occurs within the matrix (Fig. 8.3). Through a series of steps which cycle back to the re-synthesis of citric acid, a series of energy-rich NADH or FADH$_2$ (nicotinic (or flavin) adenine dinucleotide) molecules are produced, in addition to a GTP (guanine triphosphate) molecule which serves the same energy-supply role as ATP. The NADH molecules are energy rich because they serve as **electron carriers**, participating in a chemical process by which energy is transferred by the flow of electrons. **Electron transfer** occurs by **oxidation–reduction reactions** which are fundamental to

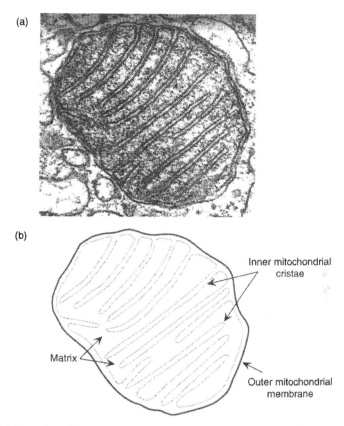

Fig. 8.3 (a) Scanning of electron micrograph of mitochondrial ultrastructure (courtesy of Hans Hoppeler). (b) Schematic diagram highlighting the outer mitochondrial membrane, the cristae formed by folding of the inner mitochondrial membrane and the mitochondrial matrix. This increases the surface area and hence the packing density of the cytochrome electron transport chains that are bound to the inner membrane surface, enhancing the rate of aerobic ATP synthesis.

the cellular metabolism of all animals and plants. Oxidation–reduction reactions involve an **electron donor** (the molecule being oxidized) and an **electron acceptor** (the molecule being reduced):

$$e^- \text{donor}(+ H^+) + NAD^+(e^- \text{acceptor}) \rightarrow NADH + \text{oxidized donor} \quad (8.4)$$

(an equivalent reaction applies to $FADH_2$). In the case of the citric acid cycle, three NADH molecules and one $FADH_2$ molecule are produced for each pyruvate that is converted into acetyl CoA from the oxidation reactions of intermediate compounds in the cycle. Therefore eight energy-rich electron carriers (six NADH and two $FADH_2$ molecules) are generated for each glucose molecule that is broken down and enters the citric acid cycle. The citric acid cycle is sustained as long as there is sufficient oxygen within the mitochondria to regenerate NAD^+ and $FADH^+$, and there is continued breakdown of glucose to form pyruvate and acetyl CoA.

The NADH and $FADH_2$ formed from the tricarboxylic acid cycle diffuse to sites within the inner mitochondrial cristae (Fig. 8.3), where they react with membrane-bound cytochrome oxidases which accept an electron from each carrier molecule in a subsequent oxidation–reduction reaction (Fig. 8.2). The flow of energy by means of electron transfer through the 'respiratory cytochrome chain' within the mitochondrial cristae ultimately fuels the aerobic production of ATP (ADP + Pi → ATP). Consequently, this process is also known as 'oxidative phosphorylation'. The final oxidation step involves combining oxygen with the H^+ that is transferred from NADH (together with the electron) to form water. Carbon released from the citric acid cycle and in the conversion of pyruvate to acetyl CoA forms carbon dioxide:

$$4H^+ + C + 2O_2 \rightarrow 2H_2O + CO_2. \tag{8.5}$$

Enough energy is released from each NADH that is oxidized to produce three ATP per NADH (two ATP per FADH2). In addition to the citric acid cycle, an additional four NADH are produced through glycolysis. Together with the net two ATP produced in glycolysis and two GTP in the citric acid cycle, this yields a total of 38 ATP molecules produced by the oxidation of one glucose molecule. Hence, 277 kcal of energy can be produced per mole of glucose that is oxidized (7.3 kcal/mol ATP). Given that a mole of glucose represents 686 kcal of energy, this means that the aerobic metabolic efficiency achieved by cells is generally about 40 per cent.

8.4.1 Fat metabolism

The breakdown of fats first involves the cleavage of the three fatty acid hydrocarbon chains attached to glycerol (Fig. 8.2). Glycerol is converted via an intermediate to pyruvate in the cytosol, while the fatty acids are transported inside the mitochondria, entering the citric acid cycle by each hydrocarbon unit being converted into acetyl CoA. Subsequent production of ATP for energy needs follows the same oxidative pathways in the mitochondria as for glucose. Owing to their larger molecular weight and higher energy content per unit weight, many more ATP are produced by the oxidation of 1 mol of stored fatty acids compared with glycogen. For example, palmitic acid (Fig. 8.1) yields a net 129 ATP, providing 940 kcal/mol of energy. However, the oxidative efficiency for burning fat is 40 per cent, similar to the efficiency of carbohydrate oxidation. Unlike plants, animals lack the metabolic enzymes necessary to convert acetyl CoA into pyruvate and hence (much as one might wish it) cannot convert fat into carbohydrate.

8.5 Quantifying energy use: respirometry measurements of oxygen consumption or carbon dioxide production

The steady state metabolism of an organism can be measured by quantifying the amount of oxygen that the organism consumes per unit time (defined as its rate of oxygen consumption $\dot{V}_{O_2}$). Alternatively, the amount of carbon dioxide that an animal produces can be measured. Both methods assume a 1:1 molar equivalence of oxygen consumption and carbon dioxide production, which is expected for glucose metabolism (eqn (8.1)). Under these conditions, the organism's respiratory quotient (RQ), defined as the ratio $\dot{V}_{CO_2}/\dot{V}_{O_2}$, is considered to be unity. However, this is not the case when substantial fat or protein metabolism occurs (the RQ is 0.71 for fat metabolism and 0.81 for protein metabolism). Consequently, the relative importance of fat oxidation to carbohydrate oxidation can be inferred by the degree to which RQ drops below unity (under most circumstances the oxidation of protein for energy supply can be assumed to be zero). Typically, between 4.7 and 5.0 kcal are produced for each liter of oxygen consumed, depending on whether fat or carbohydrate is the principal fuel. Given that 1 kcal = 4.184 kJ, a value of 20.1 kJ of ATP is commonly used for each liter of oxygen consumed (or 20.1 J/ml O_2).

Measurements of oxygen consumption or carbon dioxide production depend on an animal being at 'steady state'. This means that changes in metabolic status and activity intensity are not occurring, so that the respirometry measurements reflect a stable and uniform state of gas exchange by the animal associated with its metabolic energy demand for oxygen. Although the time-lags between the onset of activity and cellular metabolism required to meet the demand for energy (a few seconds) and between cellular metabolism and the elevation of respiratory and cardiovascular gas transport processes (a few more seconds) are short, the lag between the animal's gas exchange status and the time when this is reliably measured by the gas analyzer (device used to measure oxygen or carbon dioxide gas content) is considerably longer (many seconds to minutes). Consequently, respirometry measurements of animal metabolism require a certain period of steady activity by the animal in order to obtain reliable measurements of steady state energy use. This time-lag can be decreased by reducing the length and size of tubing connecting the chamber or mask to the gas analyzer equipment and by increasing the flow rate.

Two methods are used to measure $\dot{V}_{O_2}$ or $\dot{V}_{CO_2}$—'closed' and 'open-flow' respirometry. Closed respirometry involves placing the animal inside a sealed container (air or water, in the case of aquatic species) from which samples are taken at regular intervals. The samples are then analyzed to determine their amount of oxygen or carbon dioxide, allowing the animal's metabolism to be calculated based on changes in oxygen or carbon dioxide content over specified time periods. Alternatively, oxygen or carbon dioxide content may be monitored continuously. While this can be done using closed respirometry, it is most often done using in open-flow respirometry, providing a continuous record of energy use (accounting for the time-lag of the measurement). Because closed respirometry makes experimental manipulations of the animal more difficult, open-flow respirometry is generally preferred, especially for air-breathing animals. Open-flow respirometry involves the extraction of a continuous gas sample from a fully mixed flow of the respired gas mixture. Although open-flow respirometry can be used for both aquatic and air-breathing animals, in practice it is most often used for the latter. In the case of larger animals, a mask can be placed over or directly in front of the animal's respiratory passages. In the case of small animals, the gas can be sampled from the chamber in which the animal is moving. Vacuum pumps are used to draw the respired gas mixture from the animal, or its chamber, and to sample the gas via input to a gas analyzer so that leaks are not a problem. The gas sampled by the analyzer, using either closed or open-flow respirometry, must be dried first before determining $\dot{V}_{O_2}$ or $\dot{V}_{CO_2}$, which is measured according to the Fick equation:

$$\dot{V}_{O_2} = Q([O_{2\,atmos}] - [O_{2\,in}]) \quad \text{or} \quad \dot{V}_{CO_2} = Q([CO_{2\,in}] - [CO_{2\,atmos}]) \quad (8.6)$$

where Q is the volume flow rate of gas respired by the animal (measured by flow meters) and $[O_{2\,in}]$ and $[CO_{2\,in}]$ are the fractional concentrations of the gas sampled by the analyzer referenced to their atmospheric concentrations (accounting for the water vapor pressure) appropriate to the conditions of the experiment.

8.6 Sources and time course of energy usage during exercise

When an animal begins to move or alters its activity level, a finite lag exists between changes in the immediate demand for ATP by its muscles and the time required to supply ATP by cellular metabolism (Fig. 8.4). In order to accommodate the immediate splitting of ATP for force and movement within the myofibrils of the muscle, there is a high-energy phosphate pool within skeletal (and cardiac) muscle. In most animals, this is creatine phosphate, also known as phosphocreatine (PCr). Biochemical studies based on nuclear magnetic resonance spectroscopy have shown that PCr is in equilibrium with the regeneration of ATP from the ADP that is produced by the splitting of ATP by myosin

$$PCr + ADP \rightarrow Cr + ATP. \quad (8.7)$$

This buffers the muscle's immediate demand for ATP, allowing time for the activation of glycolysis to generate ATP anaerobically. Anaerobic production

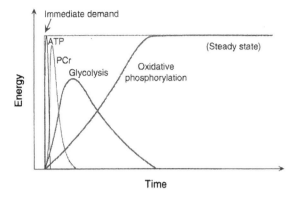

Fig. 8.4 Diagram showing how the immediate demand for an increase in ATP supply (gray line) to the muscles of an exercising animal is met by the various sources of metabolism during the start of exercise. See text for further details.

of ATP continues until oxygen delivery to the mitochondria can be increased to meet the demand for energy. Delivery of oxygen to the mitochondria depends on a multistep process which is initiated with increased oxygen uptake by the respiratory system. In vertebrates this is linked in series with bulk transport of oxygen by the cardiovascular system to the muscle cells and other tissues, and subsequently diffusion to the mitochondria. The capacity for oxygen delivery at each of these steps is fairly closely matched to the aerobic capacity of the species (Weibel *et al.* 1981). In insects, a tracheal system allows oxygen to diffuse directly to metabolizing cells. However, debate remains as to whether a pumping mechanism exists that would facilitate bulk transport via the tracheoles for oxygen delivery.

8.6.1 Oxygen deficit, post-exercise oxygen recovery and steady state metabolism

Following its initial supply from the PCr pool, ATP production by cellular metabolism is first achieved mainly through anaerobic glycolysis. This is because glycolysis can be rapidly activated, well in advance of the increased delivery of oxygen to the mitochondria (Fig. 8.4). Associated with the early glycolytic synthesis of ATP, lactate is also produced. In addition to PCr, many invertebrates also store arginine phosphate in high concentrations in their muscles. This provides these animals with an additional high-energy phosphate store which can be utilized to offset their reliance on anaerobic glycolysis, improving their endurance (see section 8.7). The relative supply of ATP via anaerobic glycolysis then begins to diminish as the animal's total demand for energy is increasingly met by aerobic synthesis of ATP within the mitochondria. As long as the energy demand for a given level of exercise is within the aerobic limit of the animal, all subsequent steady state ATP production is achieved by aerobic metabolism. At this point, the animal can be considered to be in a state of **steady state aerobic metabolism**, which is the period when an animal's aerobic metabolic rate can be reliably measured.

The initial supply of ATP from the PCr pool and through anaerobic glycolysis can be thought of as representing an 'oxygen deficit' incurred by the animal. Once the animal ceases activity and returns to a resting state, it begins to 'pay back' this deficit (Fig. 8.5(a)). As with the onset of ATP demand when exercise commences, ATP demand immediately declines to a resting level once the exercise bout ends. Nevertheless, an animal's metabolic rate remains elevated for some period of time after the end of activity, gradually declining to a resting level. This elevated post-exercise oxygen metabolism, or 'oxygen debt', reflects the need for aerobic synthesis of ATP to regenerate the PCr pool that was depleted at the onset of activity, as well as the need to re-convert lactate to pyruvate. With sufficient oxygen available, the lactate can be converted back to pyruvate, which is then oxidized by means of the tricarboxylic acid cycle and oxidative phosphorylation. It also allows other metabolic intermediates to be re-established to their pre-exercise levels. This pattern of metabolism and energy use is simple to verify. After jogging, a person's heart rate and breathing rate often remain elevated for a noticeable period of time (often referred to as 'catching one's breath'). This reflects the metabolic need for continued oxygen delivery to the mitochondria to fuel the continued aerobic production of ATP associated with 'payment' of the oxygen debt after the end of exercise.

Owing to differences in the metabolic pathways used to oxidize glucose to meet the initial demand for aerobic energy supply and the subsequent recovery phase of metabolism to re-establish resting pools of intermediate metabolites (regeneration of PCr, breakdown of lactic acid, re-synthesis of glycogen etc.), the areas under the curves representing the oxygen deficit and oxygen debt are *not* necessarily equal. Because of this, the period of elevated oxygen consumption associated with re-establishing a balanced resting metabolic state is now more commonly referred to as 'excess post-exercise oxygen consumption' (EPOC) (Gaesser and Brooks 1984). Nevertheless, for the purposes of broader comparisons of energy supply relative to exercise intensity across animals with differing metabolic strategies, as well as when comparing more athletic versus less athletic species, it is reasonable to assume that the amount of oxygen consumed during post-exercise recovery (oxygen debt) is an approximate measure of the deficit in energy supply incurred at the start of exercise (Fig. 8.5(a)).

For endothermic (birds, mammals and flying insects) and a few regionally endothermic animals (e.g. tuna) with intermediate to high aerobic capacities which allow for sustained activity over more prolonged periods of time, the oxygen deficit incurred at the onset of activity is typically a small fraction of the total amount of energy expended over the course of an exercise

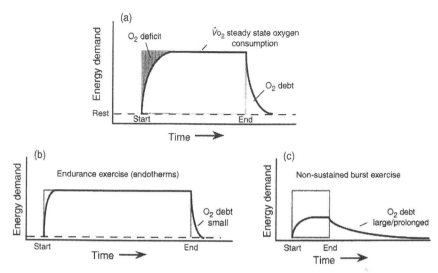

Fig. 8.5 (a) Diagram showing energy demand over the course of several minutes of steady exercise. The lag in the rate of aerobic metabolism needed to meet the energy demand of exercise results in an oxygen deficit at the onset of the exercise bout, which ends once the animal reaches a steady state level of aerobic metabolism ($\dot{V}_{O_2}$). This is 'paid back' as the oxygen debt at the end of exercise. This elevated post-exercise oxygen consumption reflects the need to re-metabolize the lactate produced by early glycolysis to meet the more immediate demand for energy, as well as to re-establish resting levels of other metabolites, including PCr (see Fig. 8.4). (b) The oxygen deficit incurred by more aerobic (and typically endothermic) species is small, allowing them to sustain exercise at higher levels for long periods of time, compared with (c) most ectothermic species which have limited capacity for sustaining exercise by aerobic energy supply and incur a large oxygen deficit, requiring them to limit their exercise to short bouts interposed with long periods of recovery metabolism.

bout (Fig. 8.5(b)). For ectothermic species which depend considerably more on anaerobic energy supply, the oxygen deficit represents a much greater fraction of their total energy expenditure. As a result, the post-exercise oxygen metabolism of ectothermic species in general is much more prolonged than that of endothermic species (Fig. 8.5(c)). As a result, ectothermic species typically utilize short bouts of exercise for foraging, escape or mating, followed by long intervals of rest and recovery metabolism.

8.6.2 Sustainable activity, maximum aerobic metabolism and metabolic scope

The ability of an animal to sustain a given level of locomotor activity requires that its metabolic demand for ATP be met by a steady state supply of oxygen for mitochondrial oxidation. Therefore an animal's maximum rate of oxygen consumption ($\dot{V}_{O_2 max}$), or **maximum aerobic capacity**, sets a limit to sustainable aerobic activity (Fig. 8.6). Technically, if an animal's demand for energy

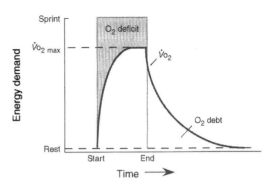

Fig. 8.6 (a) Diagram showing an exercise bout which exceeds an animal's maximum aerobic capacity or $\dot{V}_{O_2\,max}$. This incurs a large deficit which requires a long period of recovery metabolism after exercise has ended. The ratio of $\dot{V}_{O_2\,max}$ to resting metabolic rate defines an animal's aerobic scope.

does not exceed its maximum aerobic capacity, it should be able to sustain that level of exercise intensity indefinitely—or at least until its principal fuel stores (glycogen and fat) are depleted. In practice, however, an animal's maximum sustainable performance is generally achieved at about 80–95 per cent of its $\dot{V}_{O_2\,max}$ (this practical limit is based largely on data for human athletes and a few mammalian quadrupeds). The reason for this remains unclear, but is probably due to a number of factors which contribute to physiological fatigue of the animal. In studies of human runners, training not only enhances a runner's $\dot{V}_{O_2\,max}$, but it also increases his or her ability to exploit a greater range of $\dot{V}_{O_2\,max}$. This results from both physiological adaptations of a person's cardiorespiratory system, which enhance uptake and delivery of oxygen to their tissues, and improved biochemical oxidative capacity within their muscles for oxidative ATP supply. By lowering the oxygen deficit incurred at the start of exercise and limiting lactate accumulation, this probably serves as one important mechanism for improving sustainable aerobic capacity. It is also probable that, as an animal approaches its $\dot{V}_{O_2\,max}$, there is a progressive increase in anaerobic ATP synthesis in a growing number of its muscle fibers. This will cause a gradual build-up of lactate and eventual fatigue, even though the animal may not be exercising at its $\dot{V}_{O_2\,max}$.

The ratio of an animal's $\dot{V}_{O_2\,max}$ to its resting (basal) metabolic rate represents its **metabolic scope**. Aerobic species generally have evolved greater scopes than species which rely more heavily upon anaerobic energy supply. Some of the highest aerobic scopes have been measured in quadrupedal mammals: dogs have aerobic metabolic scopes of about 30, and horses and pronghorn antelope (Lindstedt *et al.* 1991) as high as 50 to 60 (i.e. their $\dot{V}_{O_2\,max}$ is 50–60 times greater than their resting metabolic rate). Recent measurements (Bundle *et al.* 1999) suggest that rheas (a flightless bird) also achieve aerobic scopes similar to dogs. In general, flying birds have aerobic scopes of about 10–20 times their resting rate. Consequently, the high aerobic scope of the ground-dwelling rhea, which is not a particularly 'athletic' animal, remains something of a mystery. Such a large factorial increase in aerobic capacity requires an extreme physiological capacity for respiratory gas exchange (ventilation rate and diffusion capacity for oxygen uptake at the lung), cardiac supply to the tissues and oxidative metabolism in the mitochondria. Athletic humans typically have aerobic scopes ranging from 20 to 30. Less active humans have aerobic scopes of 15, or even less. Generally, across a diverse range of vertebrate taxa, metabolic scopes in the range of 6 to 20 are observed. In a broader sense, therefore, humans are a fairly aerobic species. Although not many healthy humans could run a marathon, the many thousands that do attest to the endurance and aerobic capacity of fit human runners.

Even though the $\dot{V}o_{2\,max}$ of ectothermic vertebrates is considerably less than that of endothermic vertebrates, ectotherms have similar metabolic scopes (5–15) to endotherms because of their lower resting metabolic rates. On average, resting rates of ectotherms are roughly 3.5 to 4 times less than similar-sized endotherms when compared at the same body temperature (Fig. 8.7). This means that their maximum aerobic capacity is similarly less. Perhaps not surprisingly, lungless salamanders which breathe through their skin have lower aerobic scopes (1.6–3.5-fold) than lunged salamanders (3.5–7.0-fold) (Full *et al.* 1988), indicating that cutaneous gas exchange is more limiting than lung ventilation. Invertebrates which have been studied (e.g. ghost crabs and cockroaches) have similar factorial scopes to vertebrates, ranging from 6.5- to 14-fold (Full 1987). Thus the factorial capacity to elevate cellular aerobic metabolism from resting rates, appears to be generally similar across a broad diversity of animals. Differences which do occur are probably related

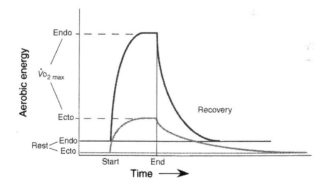

Fig. 8.7 Diagram comparing the time-course for the maximal responses of aerobic metabolism during sprint exercise in an endotherm (Endo) with that in a similar-sized ectotherm (Ecto). Although ectotherms have a much lower maximal aerobic capacity than endotherms, both groups of animals have similar aerobic scopes. This results from the 3.5–4-fold lower resting metabolic rate of ectotherms (when compared at an equivalent body temperature). Hence, in general both groups have the same capacity for elevating their aerobic metabolism with respect to resting rates. However, there is a great range of variation with respect to aerobic scope among both groups of animals. See text for further details.

to differences in the density of mitochondria within the cells and the oxygen delivery capacity of different species.

8.7 Endurance and fatigue

When an animal's demand for ATP exceeds its aerobic capacity, it must rely increasingly on anaerobic energy supply (Fig. 8.6). This results in the production of lactic acid, leading to progressive acidosis and impaired physiological function, or **metabolic fatigue**. As a result, an animal's endurance depends strongly on its aerobic capacity. Highly aerobic species have greater endurance at a given level of exercise intensity compared with less aerobic species. This is particularly the case when comparing endothermic species (flying insects, birds or mammals) with ectothermic species (most invertebrates, most fish, amphibians and reptiles). The evolution of a heightened aerobic capacity afforded by an endothermic strategy was probably important to the ability of these animals to exploit a wider thermal niche range, favoring broader daily and seasonal activity strategies, as well as the successful invasion of more diverse terrestrial climates. On the other hand, reduced rates of metabolism enable ectothermic species to achieve more economical metabolic strategies. This reduces their longer-term energy requirements, potentially improving their ability to confront environmental challenges, but limits their endurance capacity.

Thus differences in thermoregulatory strategy exert a considerable effect on the profile of energy metabolism during exercise. Endothermic species are capable of prolonged periods of sustained activity, with brief periods of recovery metabolism (Fig. 8.8(a)). In contrast, ectothermic species usually exceed their aerobic capacity for all but very slow speeds of movement, requiring much longer periods of recovery (Fig. 8.8(b)). Consequently, whereas endotherms maintain moderate to high levels of activity for long periods, ectotherms generally use brief bouts of activity when foraging, or burst activity to catch prey or avoid predators. As mentioned above, some invertebrates store arginine phosphate in their muscles to provide a buffer against the onset of glycolytic acidosis and fatigue. Nevertheless, compared with the aerobic capacity of endothermic mammals, birds and flying insects, the endurance capacity of ectothermic animals is quite limited.

Although endurance and movement distance are strongly affected by an animal's aerobic capacity, differences in aerobic capacity do not translate into differences in top speed. The burst speeds of ectotherms and endotherms are generally indistinguishable when compared at similar body temperatures and broadly across size and habitat. This reflects the fact that, owing to the same contractile proteins, the capacity of skeletal muscle to generate mechanical force and power is generally uniform across a diversity of vertebrate and invertebrate taxa when operating under similar conditions. When sufficient

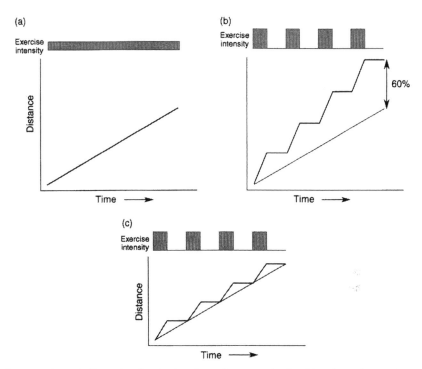

Fig. 8.8 Schematic illustration showing how intermittent exercise, involving shorter bouts of more intense activity, compared with steady exercise at a lower intensity may enable an animal to achieve greater distances before becoming fatigued. (a) Pattern for steady exercise and (b) a pattern in which exercise involves shorter and more intensive bouts (2.5-fold increase). In this example, a 40 per cent duty cycle (i.e. 60 per cent of the time is spent resting) enables the animal to go 60 per cent further than it would have done by moving at a slower, but steady, speed. However, it should be recognized that not all species necessarily benefit by intermittent exercise strategies. (c) An example in which intermittent activity does not alter the cost of transport.

ATP is available, the contractile capacities of endothermic and ectothermic muscles for work and force are broadly similar. Hence, it is the time-course of ATP supply to the muscles and the underlying metabolic pathways which produce ATP that determine the capacity for sustaining the operation of muscles at a given level of locomotor intensity. Endurance and maximum sustainable activity (or aerobic speed) are the features which vary considerably owing to differences in thermoregulatory strategy and aerobic capacity.

8.8 Intermittent exercise

Metabolic responses to exercise are nearly always studied under steady state conditions of exercise (while an animal moves at a constant speed on a treadmill, swims at a constant speed in a flow tank or flies at a constant speed in a wind tunnel). However, in nature most animals exhibit intermittent patterns of activity of varying intensity. Recent studies of intermittent exercise in both invertebrate and vertebrate runners show that such activity patterns can substantially enhance an animal's endurance or its capacity for moving longer distances, i.e. the time-averaged speed and distance over which an animal can move without fatiguing is increased by interspersing periods of rest and recovery metabolism between bouts of activity.

This is schematically illustrated in Figures 8.8(a) and 8.8 (b). In this example, an intermittent pattern of activity (work duration, 40 per cent; pause duration, 60 per cent) (Fig. 8.8(b)) provides a 60 per cent increase in endurance capacity, measured in terms of the total distance the animal travels over the time period illustrated. Over a shorter period, an animal can increase its exercise intensity without becoming fatigued. Periods of intervening rest allow the animal to recover. This allows the animal to move further over the full period of exercise and rest. In ghost crabs, intermittent patterns of activity and brief pauses have been shown to increase the distance that the animals can travel two- to fivefold, and a 1.7-fold increase in distance capacity has been observed in lizards (frog-eyed gecko) (Weinstein and Full 2000). Intermittent patterns of activity have generally been hypothesized to be more important for enhancing endurance in ectothermic species which have a low aerobic scope. However, even for endothermic species, such as humans, intermittent exercise may enhance endurance capacity. Indeed, this is the basis for interval training in track and field. Sprinters and middle-distance runners typically train by running a number of short bouts of intensive high-speed sprints interspersed by brief periods of rest, rather than running at high speed for a single longer fatiguing bout of exercise. This enables runners to achieve a longer period of intensive exercise training (at high speed), which strengthens and improves their endurance capacity for intermediate and sprint exercise.

Intermittent swimming has also been observed recently via video cameras mounted on diving weddell and elephant seals, as well as bottlenose dolphins and blue whales (Williams *et al.* 2000). These diving marine mammals must swim without breathing while diving. They increase their aerobic capacity and, as a result, the distance and depth of their dives by gliding during the descent of a dive. Intermittent periods of gliding have also been observed during ascent. The ability to glide while descending appears to result from their collapsed lungs which allow them to achieve negative buoyancy.

The ability to increase endurance by means of intermittent patterns of exercise probably involves the regular replenishment of intermediate fuel substrates before products of anaerobic glycolysis build up to metabolically disruptive and fatiguing levels. Maintaining such end-products at modest levels within the body is believed to facilitate the rate at which recovery can be achieved, readying an animal for a subsequent bout of exercise. Further studies of intermittent exercise are needed to define better the biochemical basis for enhanced endurance and for relating this to intermittent patterns of activity which nearly all animals employ under natural conditions. Hence, contrary to a favorite child's tale of the tortoise and the hare, slow and steady does not always win the race!

8.9   Other adaptations for increased aerobic capacity

In addition to intermittent exercise, various animals show specialized adaptations for increased endurance. One of the more remarkable of these is the capacity of tunas and lamnid sharks (makos and great whites) to achieve regionally elevated temperature of their swimming muscles (Carey 1973). In contrast with most fish, which have a narrow band of red muscle located near the body surface (Chapter 4), these heterothermic 'warm-muscle' fish possess an enlarged region of red muscle located deep within the myotome. Associated with this novel anatomic location, the red muscle is supplied by a vascular network, or **rete**, of arteries which pass from the outer body surface inward to the muscle. As the colder oxygenated blood carried by the arteries from the gills moves inward, it is warmed by warm venous blood passing in the opposite direction from the exercising red muscle. By the time the venous blood reaches the body surface before returning to the heart it loses nearly all of its heat to the inward-flowing arterial blood. This vascular network forms a **counter-current heat exchanger** which traps the metabolic heat produced by the active red muscle, preventing it from being lost to the surrounding cooler water. This allows the red muscle to operate at temperatures which may be up to 15 °C greater than the surrounding water. In other fish, metabolic heat produced by their swimming muscles is lost when the warmed venous blood is carried to the gills for gas exchange. Because of the red muscle's elevated temperature, tuna and other 'warm-muscle' fish can operate with increased aerobic capacity in much cooler waters than other large fish. In addition to their swimming muscles, tunas and billfish also keep their viscera, brain and eyes warm (Block 1986, 1994). This obviously has the benefit of improving their visual and mental capacity as large predators in open cooler waters.

Another adaptation for increased endurance is found in diving mammals which must carry their oxygen stores on board when they dive to depth for long periods. In addition to other physiological adaptations, elephant seals, Weddell seals, and other diving mammals possess unusually large amounts of myoglobin within their muscles, which allows them to store much more oxygen within their body than terrestrial mammals of similar size (Kooyman and Ponganis 1998). The ability to store oxygen and to avoid excessive lactate production and ensuing metabolic acidosis is clearly critical to their ability to dive for long periods before resurfacing to breathe.

8.10 Summary

The metabolic production of ATP for fueling locomotor activity, and the many other energy-demanding activities of cells, follows highly conservative biochemical pathways. This means that common principles underlie the time-course and relative dependence on aerobic versus anaerobic supply of energy in an enormously diverse range of motile animals. The relatively slow rate of oxygen delivery required to increase and achieve a steady state aerobic fuel supply necessitates the existence of high-energy phosphate stores within the cell and the early reliance on anaerobic ATP synthesis to enable an animal to be active immediately. This necessitates a period of recovery metabolism to re-synthesize these high-energy stores and return anaerobic intermediates to resting levels. The period of recovery metabolism is longer for less aerobic species and when exercise is more intense, particularly if exercise intensity exceeds an animal's maximal aerobic capacity $\dot{V}_{O_2\,max}$. Intermittent patterns of activity may increase an animal's stamina and delay metabolic fatigue. In general, the aerobic scope of most animals is in the range of 10 to 20 times an animal's resting metabolic rate; however, this can vary enormously. Highly aerobic species have evolved scopes that exceed 40–50, whereas less aerobic species have scopes as low as 2–3. The penalty of an elevated body temperature and aerobic scope is the high metabolic rate and maintenance cost that this entails. Less aerobic species can operate economically and avoid the high energy intake demands to meet their activity needs.

The evolution of increased aerobic capacity and stamina is linked to common physiological features. Increased aerobic capacity and endurance is generally associated with an elevated body temperature. This is a default thermoregulatory strategy of endothermic animals, such as birds and mammals. However, some fish have evolved the capacity to heat critical regions of their body to get around the problem of heat loss by gill respiration. Similarly, many insects have evolved the capacity to be endothermic by undergoing pre-flight warm-up to elevate the temperature of their thorax and flight musculature. Basic thermal effects on reaction rate underlie the increased aerobic capacity of these animals which allows them to sustain higher levels of activity. Even anaerobic ectothermic taxa (such as reptiles and amphibians) behaviorally warm themselves early in the day to enhance their capacity for sustainable exercise. However, in those animals that regularly can achieve moderate to high levels of aerobic activity, there must be associated physiological adaptations of the respiratory and cardiovascular systems for effective oxygen uptake and delivery to the mitochondria, sufficient numbers of mitochondria in the cells and a suite of aerobic enzymes which function effectively at warmer temperatures.

9 | Energy cost of locomotion

The locomotor activity of an animal represents one of the most important components of its energy budget. Whether an animal's movement is rapid and brief, as when a lizard escapes from a predator into its burrow, or it is more prolonged, as when a bird or an African antelope migrates many hundreds of miles, the energy cost of movement is central to the biological success of the species. In this chapter we shall first focus in detail on the energetic cost of terrestrial locomotion, examining how energy cost varies with running speed and change of gait. The effects of body size (scaling) and the animal's thermoregulatory physiology will also be considered. Patterns that emerge from studies of terrestrial locomotion, which has been best studied, will then be compared with patterns observed for swimming and flight.

Not surprisingly, the sources of metabolic energy discussed in the previous chapter underlie much of the difference in the patterns of energy use and capacity for aerobic metabolism observed among different species. Similarly, differences in physical environment underlie much of the difference in pattern observed between locomotor modes.

9.1 Energy cost versus speed of terrestrial locomotion

In order to run faster, animals move their limbs more rapidly and reduce the time that their feet remain in contact with the ground (see Chapter 3, Fig. 3.2). Because of this, an animal's muscles must generate greater forces, contract more quickly and work at greater rates when moving at a faster speed. All of these require a greater rate of metabolic energy supply. Somewhat surprisingly, the rate of metabolic energy expenditure (measured via the animal's oxygen consumption—see Chapter 8, section 8.5) increases linearly with increasing speed in a diverse range of vertebrate and invertebrate terrestrial

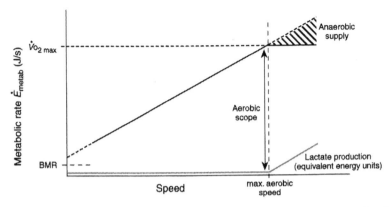

Fig. 9.1 An animal's metabolic rate $\dot{E}_{metab}$ obtained from steady state oxygen consumption $\dot{V}_{O_2}$ measurements as a function of running speed. $\dot{V}_{O_2}$ increases linearly with respect to speed in nearly all terrestrial species up to a maximal level $\dot{V}_{O_2\,max}$, which determines the maximum aerobic capacity of an animal. At greater speeds the additional demand for energy must be supplied by anerobic glycolysis, which leads to fatigue. The intercept of $\dot{V}_{O_2}$ versus speed exceeds the basal metabolic rate (BMR) of the animal. See text for additional details.

animals (Fig. 9.1). This also generally holds across changes of gait. The linear increase in energy expenditure is surprising because any increase in cost associated with the kinetic energy of swinging the limbs back and forth would suggest an exponential increase in cost with speed ($\propto v^2$). The cost due to drag resulting from air resistance on the animal's body would also be expected to increase in proportion with v^2 (it is because of this that automobiles suffer rather large increases in fuel cost at higher speeds). However, except perhaps when there is a stiff wind, most animals move too slowly or are too small for air resistance to be a significant factor. The linear increase in $\dot{V}_{O_2}$ ($\dot{E}_{metab}$) with speed observed for a wide variety of legged animals indicates that limb kinetic energy and drag are not as important as the cost associated with the *magnitude* and *rate* of muscle force generation.

The slope of the line relating the rate of energy use versus running speed (J/s divided by m/s = J/m) represents the *net* cost of transport, or the amount of energy that an animal consumes to transport itself a given distance:

$$C_{net} = (\dot{E}_{metab} - y\text{-intercept cost})/\text{speed}.$$

It is equivalent to a vehicle's fuel cost (kilometer per liter of fuel). Because of the linear (constant-slope) increase in the rate of energy use with speed, the net cost of transport for a running animal is the same at any speed. This is a rather remarkable result. It means that the amount of energy that an animal uses to run 1 km is almost the same whether it runs very fast or at a leisurely pace. We shall explore the basis for this observation below (section 9.2.1). Note, however, that this is distinct from the intermittent patterns of locomotion discussed in the previous chapter, in which bouts of exercise are often interspersed by periods of rest.

When the line relating an animal's oxygen consumption rate to running speed is extrapolated to zero velocity, the y-intercept of the line is typically about 70 per cent greater than the animal's actual resting oxygen consumption. This y-intercept can be considered a 'start-up' cost, but its basis is not well understood. If this start-up cost is included, together with the incremental increase in energy cost with speed, the total cost of transport of the animal can be calculated ($C_{tot} = \dot{E}_{metab}/$speed). Therefore an animal's total cost of transport is greater at slower speeds because the y-intercept represents a larger fraction of the total energy cost than it does at higher speeds.

As an animal runs faster, its rate of oxygen consumption eventually levels off reaching some maximal level ($\dot{V}_{O2\,max}$). The speed at which an animal attains its $\dot{V}_{O2\,max}$ defines its maximum aerobic speed (MAS). Theoretically, this sets the limit for an animal's sustainable activity. Above this speed, the animal's oxygen consumption remains fixed at a maximal rate. The additional energy required to move at faster speeds must be met by anaerobic ATP supply. Because anaerobic metabolism results in the rapid accumulation of lactate (or other anaerobic end-products in the case of invertebrate runners), movement above an animal's MAS quickly results in metabolic acidosis and fatigue. Sprint or intensive exercise of this nature requires that an animal becomes inactive to allow it to recover from the lactate accumulated within its tissues, as well as the need to replenish its fuel reserves and re-establish resting levels of intermediate metabolites (see Chapter 8, sections 8.6 and 8.7). The combination of an increase in blood lactate levels and a leveling off in oxygen consumption are the markers that signal an animal has reached its $\dot{V}_{O2\,max}$.

Although ideally an animal should be able to sustain running speeds up to its MAS, in actuality sustainable exercise is generally limited to about 80–85 per cent of $\dot{V}_{O2\,max}$. This is probably due to imbalances in metabolic intermediates which arise during submaximal, yet intensive, exercise. Even though exercise is below the animal's $\dot{V}_{O2\,max}$, such imbalances are believed to trigger sufficient anaerobic glycolysis and lactate build-up so that fatigue eventually ensues. Not only can endurance exercise training in human athletes elevate both their $\dot{V}_{O2\,max}$ and MAS, it can also enhance the fraction of $\dot{V}_{O2\,max}$ over which sustainable performance can be achieved (typically as high as 95 per cent of $\dot{V}_{O2\,max}$).

9.1.1 Intermittent activity and energy cost

Both the net and total costs of transport calculated as described above focus on an animal's steady state energy use and assume that the anaerobic start-up costs ('oxygen deficit' discussed in Chapter 8) equal the pay-off costs ('oxygen debt') of aerobic recovery metabolism at the end of exercise. However, if the post-exercise oxygen consumption differs from the metabolic cost required to begin exercise, this can alter the total metabolic cost of transport for an animal and so may be of special significance for the intermittent patterns of activity which many animals use under natural conditions (Baker and Gleeson 1999). Short intensive bouts of exercise have been shown to elevate transport costs in mice substantially (determined as the energy consumed during exercise plus that consumed during recovery divided by distance). In humans, post-exercise recovery metabolism increases with both exercise intensity and duration. However, establishing a return to resting metabolism after a period of recovery often makes an accurate assessment of the post-exercise recovery metabolism difficult. In mice, the post-exercise recovery metabolism was as great as 90 per cent of the energy consumed during a minute of intense exercise. This suggests that post-exercise recovery metabolism may constitute a considerable cost to animals. However, if the post-exercise cost of metabolism is equivalent or close to the anaerobic costs incurred at the start of exercise (equal areas shown in Fig. 8.5), it is likely that the measurement of a steady state net cost, as described above, and that based on summing the energy consumed during exercise and during recovery and dividing by distance should be similar. A subsequent study (Edwards and Gleeson 2001) found that repeated bouts of exercise at frequent intervals (13 times at 15-s intervals) resulted in a similar transport cost to steady movement over the same distance. The manner in which post-exercise energy metabolism is affected by differences in exercise intensity and duration has mainly been examined in humans and mice. More studies are needed to assess whether recovery metabolism differs substantially from start-up anaerobic costs and whether this significantly impacts the cost of transport of animals under natural conditions of intermittent activity.

9.1.2 Energy cost versus speed during wallaby and kangaroo hopping

In contrast with the linear increase in the rate of aerobic energy use versus speed that nearly all terrestrial animals exhibit, kangaroos and wallabies are two groups of moderate to large marsupials whose rate of energy use levels off when they hop and, distinct from all other animals studied, does not increase at faster speeds (Dawson and Taylor 1973; Baudinette et al. 1992) (Fig. 9.2). The steep increase at slow speeds reflects their pentapedal 'walking' gait which involves the use of their large tail in combination with their fore and hind limbs. This works well for foraging, but the animals hop to move over greater distances. The ability to hop at greater speed with no increase in energy expenditure rate mainly reflects the increased elastic

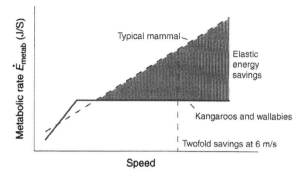

Fig. 9.2 Once a kangaroo or wallaby begins to hop its rate of energy expenditure does not increase at faster speeds, in contrast with the increase observed in other terrestrial animals. This difference probably reflects the ability of these animals to store and recover substantial elastic energy in their long leg tendons. The steep increase at low speeds reflects their less economical pentapedal walking gait.

energy savings achieved in their distal leg tendons (discussed in Chapter 3). Moving at faster speeds necessarily involves greater muscle force but not necessarily greater muscle work. The energy cost associated with more forceful and faster muscle contractions is probably offset by the increased strain energy recovered from these animals' tendons. Wallabies and kangaroos also hop faster by increasing their stride length rather than altering their hopping frequency. This may help to slow the rate of muscle force development at faster speeds, favoring a more uniform metabolic rate (see section 9.2.1). The economical hopping ability of kangaroos probably reflects their need to forage over wide areas to exploit the sparse vegetational food resource which characterizes much of their arid habitat in Australia. In contrast, the rate of energy use in smaller hopping animals (including both marsupials and rodents) increases with hopping speed similar to other terrestrial animals of their size. This reflects, in part, the inability of these smaller animals to achieve significant energy savings due to their relatively thick tendons (see Chapter 3, section 3.11).

9.1.3 Energy cost versus gait

For most animals that have been studied (mainly endothermic birds and mammals) there is little evidence of a change in metabolic cost as a function of speed within and across changes of gait. Two notable exceptions are humans and horses. For humans (Fig. 9.3), the net metabolic cost of running is as much as 50–100 per cent greater than that of walking and does not change appreciably over a range of aerobic running speeds (Margaria 1976). On the other hand, the cost of transport for walking varies with speed and has a minimum at about 1.3–1.5 m/s, indicating an optimal speed for minimizing energy use. The most likely explanation for the decline in cost from both slow and fast to moderate

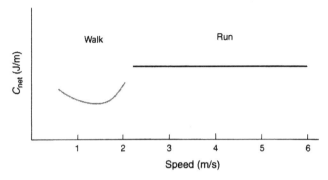

Fig. 9.3 The net energy cost of transport (J/m) versus speed is curvilinear when humans walk, with a minimum cost at an intermediate walking speed of about 1.3–1.5 m/s. The net cost of running is about 50 per cent greater than during walking but remains fairly constant across running speeds.

walking speeds is that the exchange between potential and kinetic energy of the body's center of mass (see Chapter 3) is maximal at intermediate walking speeds. In humans and some avian bipeds maximal exchanges of 70 per cent have been observed during walking. At faster walking speeds, the efficiency of this exchange declines.

The reason for the sharp increase in cost of transport from walking to running is less certain; however, two factors are likely important. First, when changing gait from a walk to a run the time t_c of limb contact with the ground drops, indicating that the muscles of the limb must generate force at faster rates. This can only be achieved by recruiting faster muscle fibers, which consume energy more rapidly (Fig. 9.4(a)). In addition, recent results indicate that the limb mechanical advantage in humans decreases sharply from walking to running (Fig. 9.4(b)), due mainly to a shift in knee posture during running. Whereas the limb is more extended during walking (reflecting the inverted pendular mechanics of this gait), the knee is more flexed during running. This helps to absorb the impact of the body during each foot fall. However, it also requires an increase in the amount of force that must be generated by the knee extensors (quadriceps). Together with the decrease in limb contact time, this probably contributes to the observed increase in metabolic transport cost during human running. More work per unit distance is also performed to move the body during running than during walking; however, there is little increase from fast walking to slow running speeds and, in contrast with metabolic cost, work per unit distance does not level off at faster running speeds (Cavagna and Kaneko 1977).

Careful measurements of oxygen consumption obtained from horses trained to extend their gaits beyond their normal speed ranges (Hoyt and Taylor 1981) have also shown that the increase in energy cost with speed is not a simple

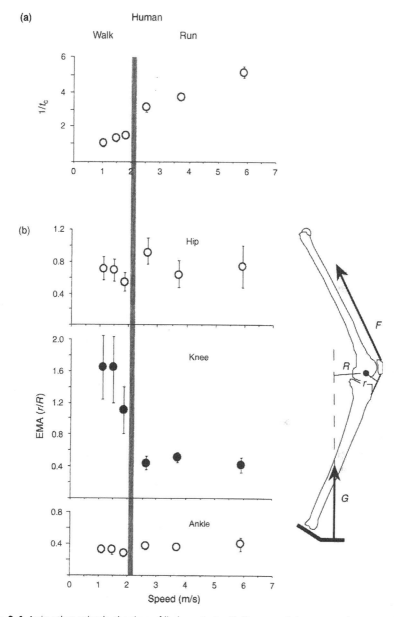

(a)

Human

Walk   Run

$1/t_c$

(b)

EMA (r/R)

Hip

Knee

Ankle

Speed (m/s)

Fig. 9.4 As in other animals, the time of limb contact with the ground decreases as humans move at faster speeds, so that (a) $1/t_c$ increases from a walk to a run. (b) The effective mechanical advantage EMA $= r/R$ of human limb extensors remains fairly uniform at different speeds within a gait, but decreases about threefold at the knee when humans change from a walk to a run. The decrease in knee extensor EMA (which results from a more flexed running posture at the knee) and the increase in $1/t_c$ probably underlie much of the increase in the energy cost of running versus walking (Fig. 9.3).

linear trend in this species (Fig. 9.5(a)). Instead, metabolic rate increases when a horse extends its speed beyond its normal gait range (e.g. when a horse trots at a slower speed that it would normally walk, or when it walks at a faster speed that it would normally trot). Consequently, when examined carefully, the metabolic cost of transport in horses is not uniform within and between gaits. A clear minimum cost of transport is observed within both a walk and a trot (Fig. 9.5(b)). By changing gait, horses are able to reduce their cost of transport so that it is maintained the same over a broad range of speeds. The linear increase in metabolic rate with running speed observed generally for other animals (Fig. 9.1), and the uniform net cost of transport that this provides, suggest that change of gait in other species may serve a similar role as it does in horses. However, because of the difficulty in documenting a curvilinear relationship between metabolic energy use and speed, this has not yet been confirmed for any other species.

When animals are allowed to select the speed at which they move, they commonly move at an intermediate speed within a gait. In the case of horses (Fig. 9.5(c)), and for humans walking, this is closely associated with their minimum cost speed and is often referred to as an animal's 'preferred speed' within a gait. The basis for the minimum cost at preferred speed observed within a particular gait for horses is not yet clear. It may reflect more effective energy conservation, such as the exchange between kinetic and potential energy in human walking, and less muscle work. Nevertheless, it seems clear that animals prefer to use a relatively narrow range of speed within a gait when moving over longer distances. Pennycuick (1975) observed similar patterns of a narrow speed range used by migrating gnu and wildebeest moving long distances over African plains. Studies of rodents and other small animals demonstrate that preferred speeds within gaits also occur over short distances. These observations suggest that, in addition to allowing animals to run faster, changes in gait can also have important energetic consequences.

9.2 Energy cost versus body size

It is obvious that larger animals expend more energy to move than small animals simply because of their greater size. But does the energetic cost of locomotion vary directly in proportion to an animal's size or weight? In order to answer this question and to compare animals of different size, the metabolic cost of locomotion can be normalized to the body mass of each species. This provides a measure of the animal's *mass-specific* rate of oxygen consumption (e.g. ml O_2/s/kg). This can be converted to J/s/kg or W/kg, assuming a value of 20.1 J/ml O_2 of oxygen consumed (converting to W/N gives an animal's *weight-specific* metabolic rate). When normalized in this way, smaller animals are observed to have higher mass-specific costs of transport

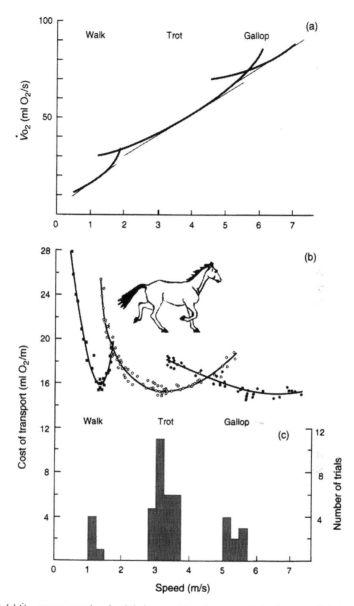

Fig. 9.5 (a) $\dot{V}_{O_2}$ versus speed and gait in horses. When horses are trained to extend their gaits, $\dot{V}_{O_2}$ increases in a curvilinear fashion with speed. (b) Because of the curvilinear increase in energy use with speed, the cost of transport in horses has a minimum at a particular speed within each gait. (c) The minimum cost of transport speeds are those that horses prefer to use when they move freely over ground using each gait. (Adapted from Hoyt and Taylor (1981), with permission from MacMillan Magazines Ltd.)

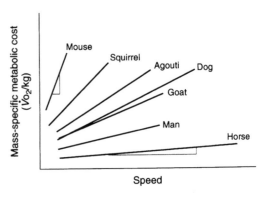

Fig. 9.6 Mass-specific metabolic energy use ($\dot{V}_{O_2}$/kg) versus speed for different-sized mammals. Small animals have a steeper increase in $\dot{V}_{O_2}$/kg with speed than larger animals and, as a result, have a higher cost of transport.

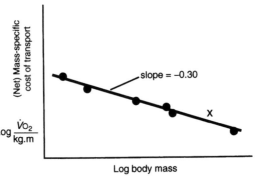

Fig. 9.7 Scaling of *net* mass-specific cost of transport (V_{O_2}/kg/m) for terrestrial locomotion as a function of body mass plotted using logarithmic coordinates. The steeper increase in $\dot{V}_{O_2}$ with speed for smaller animals (Fig. 9.6) reflects their higher mass-specific cost of transport, resulting in a negative slope (−0.30). This indicates a scaling relationship of net cost/kg $\propto M^{-0.30}$. The "X" *value* is the net cost of human runners, which is slightly higher than a mammalian quadruped of similar size. This may be linked to a human runner's lower overall limb EMA (Fig. 9.4(b)).

than large animals. This means that slopes of mass-specific oxygen consumption versus running speed vary inversely with body size (Fig. 9.6). For example, it costs a mouse 20 times more energy to move a given gram of its mass a unit distance than a dog and 30 times more energy than a horse. When compared over a range of sizes, the scaling of mass-specific cost of transport in terrestrial animals has been found to obey the very general relationship (Fig. 9.7)

$$C_{net}/kg \propto M^{-0.30}.$$

This relationship appears to apply equally well to both endothermic and ectothermic vertebrate runners (Taylor *et al.* 1982), as well as to invertebrate runners (Full 1991).

9.2.1   Rate of muscle force development and energy cost

The size- and speed-related differences in energy cost were originally thought to be due to differences in the amount of work an animal's muscles perform to transport its weight and to move its limbs. However, in a series of studies Heglund *et al.* (1982) showed that this is not the case. Instead, they found that different-sized animals perform the same mass-specific work per unit time to move their centers of mass and their limbs. In addition, the rate of work performed does not increase with speed similar to the increase in metabolic rate.

Instead, the basis for both the increased rate of energy use with speed and the scaling of terrestrial locomotor energy cost versus size appears to be largely determined by differences in the *rate* and *magnitude* of force that the limb muscles of terrestrial animals generate to support their weight while running. This largely depends on changes in the stride frequency and step length of different-sized animals as they move at different speeds (Heglund and Taylor 1988). As discussed in Chapter 3, animals move their limbs more rapidly, increasing their stride frequency (as well as their stride length) in order to increase speed. This requires that their limb muscles contract and develop force at faster rates. However, for an animal of a given size, the *time-averaged* force that its muscles must generate to support its body weight remains constant, irrespective of the speed at which it runs. This reflects the fact that increases in force magnitude are offset by the reduction in limb support time. Consequently, this suggests that the increased rate of energy use to move at faster speeds probaby results from the increased rate of force development that is required to operate an animal's muscles. This is because faster rates require more costly biochemical energy turnover within the muscle's cells.

The rate of muscle force development has been estimated to vary with the inverse of limb contact time with the ground ($1/t_c$). In a study of different-sized mammals, Kram and Taylor (1990) observed that the weight-specific rate of metabolic energy expenditure (J/s/kg or W/N) varied inversely with ground contact time t_c, or directly with $1/t_c$ (Figs 9.8(a) and 9.8(b)). The match between the rate of energy expenditure and $1/t_c$ suggests that there is a constant relationship between the amount of energy expended per Newton of body weight (J/N) that is supported on the ground by the limb (Fig. 9.8(c)). Kram and Taylor (1990) termed this the **cost coefficient** c which relates the weight-specific rate of energy consumption and rate of ground force application:

$$\dot{E}_{metab}/W_b = c(1/t_c).$$ (9.1)

Equation (9.1) indicates that terrestrial mammals and likely other animals consume the same amount of energy per unit weight for each step that they take. Therefore the use of faster-contracting muscle fibers appears to underlie

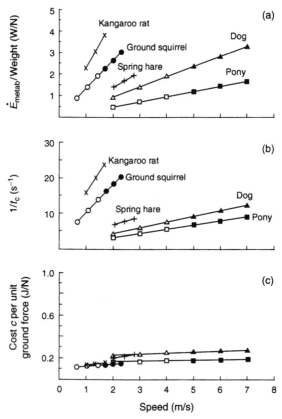

Fig. 9.8 (a) The rate of energy used per unit body weight ($\dot{E}_{metab}/W$) parallels the increase in $1/t_c$ as a function of speed for different-sized running and hopping animals. (b) When energy use is normalized to the force G that an animal exerts on the ground (E_{metab}/G (J/N)) all animals are found to consume the same amount of energy per unit ground force, irrespective of the speed that they run. This cost coefficient c is about 0.2 J/N. See text for additional details. (Adapted from Kram and Taylor (1991), with permission from MacMillan Magazines Ltd.)

much of the observed increase in metabolic energy expenditure with increasing speed. To the extent that limb muscles must also shorten to perform mechanical work, part of the increase in energy cost may also reflect the need to recruit more muscle to support body weight at faster speeds because of the force–velocity effects of the skeletal muscles. Recall (Chapter 2) that, by shortening at a faster rate, the force that a muscle can exert decreases. Thus more fibers must be recruited to maintain a given force, which increases the muscle's rate of energy use.

In addition to the effect of the rate of force development on energy cost, larger animals use lower stride frequencies and have longer limbs, allowing them to

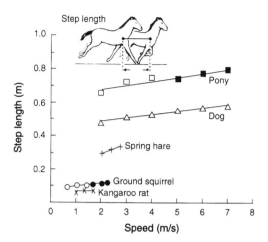

Fig. 9.9 Step length as a function of speed in different-sized animals. The lower cost of transport of larger animals is explained by the fact that they move a greater distance during the time that their limbs are in contact with the ground (greater step length) but consume proportionally the same amount of energy to support their weight (constant c; see Fig. 9.8(c)). See text for additional details. (Adapted from Kram and Taylor (1991), with permission from MacMillan Magazines Ltd.)

cover a greater distance during each stride compared with small animals. An animal's running velocity v can be represented by its step length L_c (the distance that it travels while each limb is in contact with the ground) divided by its limb contact time:

$$v = L_c/t_c. \tag{9.2}$$

This allows the mass-specific metabolic cost of transport ($C_{tot} = \dot{E}_{metab} W_b^{-1} v^{-1}$) to be defined as

$$C_{tot} = c(1/L_c). \tag{9.3}$$

Because the amount of energy that different-sized animals consume per unit weight is the same during each step, larger animals are able use this energy to move a greater distance (Fig. 9.9). The longer step (and stride) lengths of larger animals provides a simple explanation for their lower mass-specific cost of transport. It is important to note that Kram and Taylor's 'force hypothesis' and the observation of a constant cost coefficient c assumes that the muscles of different-sized animals operate over similar regions of their force–velocity curve and perform proportionately similar work at equivalent running speeds (Heglund *et al.* 1982; Taylor 1994). This may be a reasonable hypothesis, but it is one that will be important to test further.

The patterns of limb contact time, stride frequency, step length and metabolism described above have been most extensively studied and established for terrestrial birds and mammals. Future studies of a broader variety of endothermic and ectothermic vertebrate and invertebrate runners will be needed to see if they show similar basic patterns.

9.3 Ectothermic versus endothermic energy patterns

In the previous chapter we discussed how differences in an animal's thermo-regulatory strategy affect the time course and relative importance of aerobic versus anaerobic sources of energy supply to its exercising muscles. What effect does this have on ectothermic versus endothermic locomotor energetics? If the locomotor energetics of a mammal and a reptile of similar size are compared, three things stand out (Fig. 9.10). First, at a comparable body temperature (37 °C), a reptile expends less energy than a mammal at any given speed. Second, the speed range over which a reptile can sustain its activity and meet its energy demands by aerobic metabolism is extremely limited compared with a mammal. Third, the rate of increase in energy metabolism with speed (slope) is the same for endotherms and ectotherms. Because an ectotherm has a fourfold lower intercept, its cost of transport is less than that for a similar-sized mammal at any given speed. The lower metabolic rate, cost of transport and lower maximum aerobic speed of ectotherms reflect the evolution of a more economical strategy. However, it comes at the price of a more limited capacity for sustainable activity. Interestingly, the slope of the line relating metabolic energy expenditure to speed is the same for similar-sized reptiles and mammals. This suggests that similar mechanisms underlie the change in metabolic rate versus speed for these two groups of animals.

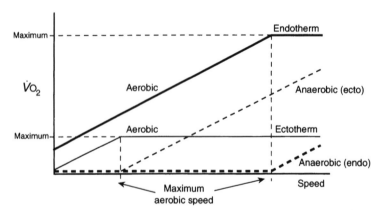

Fig. 9.10 A comparison of the rate of energy use $\dot{V}_{O_2}$ with the running speeds of a lizard and a mammal of similar size. Both animals exhibit the same increase in $\dot{V}_{O_2}$ with speed, but the aerobic capacity $\dot{V}_{O_2\,max}$ of the ectothermic lizard is much more limited than that of the endothermic mammal. This greatly limits its maximum aerobic speed. Consequently, lizards and most other ectotherms rely on short bursts of activity fueled by anaerobic glycolysis. This requires a long period of recovery metabolism and rest between exercise bouts. At any given aerobic speed, the rate of energy use by the lizard is less than that of the mammal.

Compared with the many studies of birds and mammals, fewer studies have been carried out on steady exercise in amphibians and reptiles. Partly, this reflects the limited aerobic capacity for sustainable exercise of most ectothermic species. Studies of salamanders, toads and other lizards that have been carried out (Bennett 1978) suggest a similar pattern of energy use versus sustainable speed as that observed for monitor lizards (*Varanus*). These lizards were the first animals used to compare reptilian locomotor energetics with those of mammals. Nevertheless, there is a need to expand studies of ectothermic vertebrate and invertebrate taxa to test how patterns of limb contact time, stride length and muscle force relate to energy use in a broader range of terrestrial animals.

9.4 Energy cost of incline running

Many animals must regularly scale rocks, trees and variable slopes in their environment. These inclines probably affect the metabolic cost of running because energy must be spent not only in moving the animal's body horizontally, but also in raising (and lowering) the animal's center of mass. The rate of mechanical work associated with increasing the body's potential energy is

$$P_{PE} = mgv \sin \beta \tag{9.4}$$

where m is the animal's body mass, g is gravitational acceleration, v is the animal's velocity and β is the angle of ascent. Equation (9.4) also applies to descent, in which the rate of potential energy loss depends on running speed and angle of descent. But most past work has focused on the energy cost of uphill running. The total cost of incline running can be considered as the sum of the cost C_{hor} of horizontal running and the metabolic cost C_{PE} of raising the body's center of mass.

If it is assumed that the efficiency of doing work is the same for the muscles of both small and large animals, the mass-specific metabolic cost of doing potential energy work should also be the same. This cost has been estimated to be the difference between the cost when running up an incline and the cost when running at the same speed on a level

$$C_{incl} - C_{hor} = C_{PE} \sin \beta. \tag{9.5}$$

Based on studies of different animals, mass-specific values from 16 to 27 J/kg/m have been proposed for C_{PE}. Because of this, the metabolic cost of incline running should be greater at a given incline for a large animal because it has a lower mass-specific metabolic cost for horizontal running than a small animal (Fig. 9.6). That is, the cost of raising the body's center of mass when running up an incline can be expected to be a greater fraction of the horizontal cost of locomotion in larger animals.

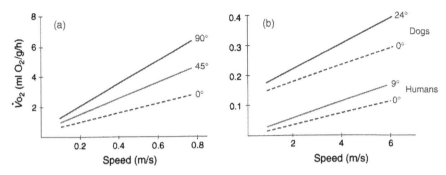

Fig. 9.11 Metabolic cost of incline running in (a) cockroaches and (b) dogs and humans. In all three species the metabolic cost of incline running is greater than the cost incurred when running horizontally. However, the extra cost cannot be explained simply by a uniform muscle efficiency and the rate of increase in potential energy work. See text for additional details.

Consistent with the extra cost of potential energy work, the metabolic cost of incline running is greater for most species that have been investigated. Figure 9.11(a) shows the patterns of oxygen consumption measured for cockroaches running on a level and up inclines of 45° and 90°. Cockroaches and other insects can not only scale vertical walls but can also run upside down. Many lizards are also able to scale vertical walls seemingly with ease. However, when cockroaches run up inclines of 45° and 90°, their metabolic cost increases twofold and threefold compared with running on a level surface. Running up an incline incurs a significant cost even for these small animals. When humans and dogs run up an incline, their cost also increases (50 per cent at 9° and 79 per cent at 24°). When studies from other different-sized animals are compared, no regular pattern for energy cost of incline running is observed (Full and Tullis 1990). If a given value for the metabolic cost of raising the body's center of mass is used (assuming constant muscle efficiency), the predicted cost of incline running is typically much greater than the observed cost (Full and Tullis 1990). In some species, it is less. Consequently, it seems likely that either muscles in different-sized animals do not operate with similar efficiency or other factors influence the energy cost of incline running. Because changes in stride frequency and ground contact time t_c are not significantly affected by incline in those species that have been examined, the cost of incline running does not appear to be affected by the rate of muscle force development, as is the case when running on a level (Fig. 9.8).

Another possibility is that postural changes associated with incline running may affect limb mechanical advantage in such a way that the magnitude of muscle force required to elevate the body when running uphill is greater than that required when running on a level. This would probably affect the amount of energy required to move uphill. Finally, increases in muscle shortening for positive work to elevate the body during incline running may increase the amount of energy that the muscles consume independent of their efficiency for doing work.

In contrast with running uphill, the energetics of downhill running has been much less studied. Except for studies of humans running on different inclines and declines (Margaria 1976; Minetti *et al.* 1994), the question of energy cost during locomotion on downhill inclines has largely been overlooked. When humans run and walk downhill, their metabolic cost decreases with a steeper angle of descent, but only up to $-10°$ (Minetti *et al.* 1994). At $-10°$, the cost is 35 per cent less than when running on a level. At steeper descent angles, the energy cost of locomotion begins to increase. During downhill running the limb muscles must increase the degree to which they contract eccentrically (active lengthening) to absorb energy. This allows for the recruitment of fewer fibers to generate a given level of force (Chapter 2) and probably contributes to the reduced locomotor cost. However, the extent to which muscles can contract eccentrically and avoid damage is limited, and this may reflect the increased cost of locomotion at very steep descent angles. Changes in stride frequency (lower when running downhill than uphill) suggest that some of the decreased cost may also reflect a slower rate of force development associated with taking longer strides. Nevertheless, the manner in which energy cost changes with downhill running in other animals, and as a function of size and gait, deserves greater scrutiny before broader generalizations of locomotor biomechanics can be established with respect to downhill energetics.

9.5   Cost of swimming

The metabolic energy cost of swimming has been measured in a diversity of vertebrate taxa, including fish, cetaceans, humans, muskrats, platypus, ducks, penguins, turtles and marine iguanas. In contrast, few invertebrate taxa have been studied except for shrimp and squid (reviewed by Videler (1993)). These measurements are based on respirometry studies of oxygen consumption made in a closed flow tank for gill breathers or with the use of a breathing chamber positioned at the water's surface for aerial breathers. These data show that, in contrast with terrestrial animals, the metabolic cost of swimming increases exponentially with speed (Fig. 9.12). This reflects the increased drag that a swimming animal experiences at higher speeds, which increases proportionally to velocity squared. The data for swimming salmon fit this more closely than those more recently reported for tuna, in which the exponent is closer to 1.5. Similar to terrestrial animals, the mass-specific cost of transport of swimming (aerobic energy cost to move a given distance and normalized for size) decreases with increasing size in fish ($\propto M^{-0.30}$). However, surface swimmers incur a three- to fivefold greater cost than underwater swimmers.

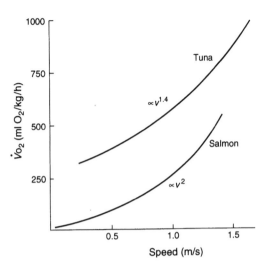

Fig. 9.12 The rate of energy use $\dot{V}_{O_2}$ of salmon and tuna increases in an exponential fashion with increased swimming speed. This reflects the increasing effect of drag ($\propto v^2$) which must be overcome at a higher speed. The exponent for salmon fits this fairly well but is less for tuna ($\propto v^{1.4}$). (Data from Brett (1965) and Dewar and Graham (1994)).

As was noted in Chapter 4, this probably results from the additional cost of wave-drag. Human swimmers have a particularly high cost of transport for their size. So, despite the fact that elite swimmers appear to be near their theoretical limit in terms of performance, it comes at a high price. Quite clearly our terrestrial locomotor capacity greatly outstrips our abilities in water. Compared with other underwater swimmers, the jetting of squid also incurs a fivefold greater cost. It is probably this heightened cost that has limited the evolution of jetting as a more common means of aquatic movement. Finally, there is evidence that fish have the lowest cost of transport for their size, with shrimp, aquatic reptiles and cetaceans having higher mass-specific transport costs (see section 9.7).

9.6   Cost of flight

As for swimming animals, the increased power required to overcome drag ($\propto v^2$) and move at a faster speed means that the metabolic rates of flying birds, bats and insects are expected to increase exponentially with speed. On the other hand, the high aerodynamic induced power cost to support body weight at slow speeds (Chapter 5) suggests that metabolic power should also rise at slow speeds, reaching a maximum during hovering. Consequently, in contrast with swimming and running, flying animals incur a high energy cost to move at slow speeds and when hovering. The decrease in induced power relative to the increase in parasite and profile power (which both result from drag) predicts a U-shaped power curve (see Chapter 5, Fig. 5.4). Energy curves of this shape suggest that there is a minimum cost speed (where metabolic energy expenditure is a minimum during flight) and a minimum cost of transport speed, also referred to as the maximum range speed (see Chapter 5, Fig. 5.4). This is determined by the tangent to the curve, which gives the lowest slope (energy rate/speed = transport cost). Birds are believed to use this speed when flying over longer distances, such as during migration.

Although the U-shaped power curve has considerable appeal and fits the pattern of energy use for fixed-wing aircraft, the evidence for U-shaped metabolic power curves in flying animals has met with mixed results. Measurements of the metabolic rates of bumblebees (Ellington *et al.* 1990) and hummingbirds (Berger 1985) during forward flight show a flat power curve over a range of speeds (Fig. 9.13(a)). The metabolic cost of hovering for these species is also the same. This remarkable result reflects the fact that bumblebees

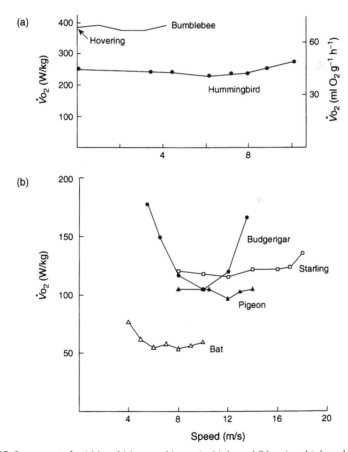

Fig. 9.13 Energy costs for (a) bumblebees and hummingbirds, and (b) various birds and bats in relation to speed, based on oxygen consumption measurements obtained from animals trained to fly in wind tunnels. (Adapted from Ellington (1991).)

and hummingbirds operate their wings (and underlying energy demands of their flight muscles) at a uniform frequency and amplitude. Changes in forward flight speed arise mainly from changes in the animal's body pitch and/or wing stroke angle, which alters the relative amount of thrust that a bumblebee or a hummingbird achieves from the lift generated by its wings. In order to fly faster, both species simply tilt forward and incline the angle of attack of their wings to produce greater thrust. As a result, their metabolic rate during flight at fast, moderate and slow speeds remains essentially the same.

Measurements of metabolic rate during flight in birds and bats have also yielded results which suggest a flatter power curve than that predicted by classical aerodynamic theory (Fig. 9.13(b)). Except for budgerigars (Tucker 1968), changes in metabolic rate with sustainable speed have not been found to exhibit a clear decline at moderate speeds and a subsequent increase at faster flight speeds. In part, this probably reflects the fact that oxygen consumption data are quite difficult to obtain for these more demanding flight speeds, and thus are not available over the full speed range of a flying animal. Except for a few bats and birds—most notably hummingbirds—the power costs during hovering are probably non-sustainable, requiring significant anaerobic energy supply. Similarly, very fast (sprinting) flight speeds are probably also non-sustainable. Consequently, flying animals probably have a narrower range of sustainable speeds than running animals. Birds also alter their flight behavior (active flapping versus gliding or bounding phases of flight, and changes in angle of attack and wing amplitude) and adjust their wing shape, which may enable them to fly at different speeds with less variation in energy cost. Finally, because flight muscles must perform considerable work to move the wing and generate aerodynamic power (see Chapter 5, section 5.5), changes in muscle efficiency which result from adjustments in muscle shortening rate and contractile behavior may also help to offset increases in energy expenditure which might otherwise occur at slower and faster speeds.

9.7 Locomotor costs compared

Compared with other forms of locomotion, swimming appears to be the cheapest means of movement (Fig. 9.14). Flying has an intermediate transport cost, and running is most expensive. The low cost of swimming may seem counterintuitive because of drag, but the buoyancy of aquatic animals reduces their need to expend energy for body support. In addition, most swimmers travel at low speeds (e.g. fish, 0.1–3 m/s) compared with flying animals (5–25 m/s), which greatly reduces drag and hence their swimming cost.

The differences in cost of transport depicted in Fig. 9.14 are based on a comparison of animals which are specialized for each locomotor mode and

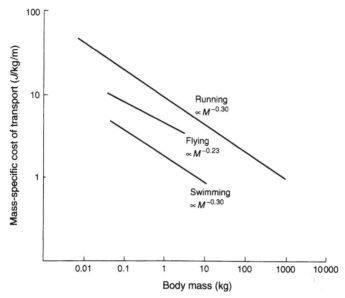

Fig. 9.14 Comparison of mass-specific cost of transport as a function of body mass for running, flying and swimming. (Adapted from Schmidt-Nielsen (1972).)

represent diverse phylogenetic groups. When more closely related animals are compared, do similar differences in transport cost emerge? Mammals represent one such group of animals which exhibit specialized terrestrial, aquatic and flying forms. Their relatively 'recent' re-invasion of the oceans 60 million years ago reflects certain differences in their swimming specialization. When marine mammal swimming specialists are compared with semi-aquatic mammals (including humans), their transport costs are much lower but are still greater than for fish (Fig. 9.15(a)). When marine mammals (dolphins, seals, whales) are compared with running mammals, the allometric lines depicting changes in cost of transport relative to body size closely overlap (Fig. 9.15(b)) (Williams 1999). The transport costs for flying bats also fall close to those for swimming and running specialists of similar size. Hence the re-invasion of an aquatic habitat by mammals probably required an initially high locomotor energy cost that was subsequently reduced with the evolution of a more specialized swimming capability (Williams 1999).

The higher transport costs of specialized marine mammals compared with fish probably reflects a higher maintenance metabolic cost in marine mammals. This is principally due to their endothermic metabolic strategy. The high thermal conductivity of water promotes a much higher rate of heat loss in water than in air. Ectothermic fish which maintain low temperature gradients between the water and their bodies do not suffer this heat loss and hence

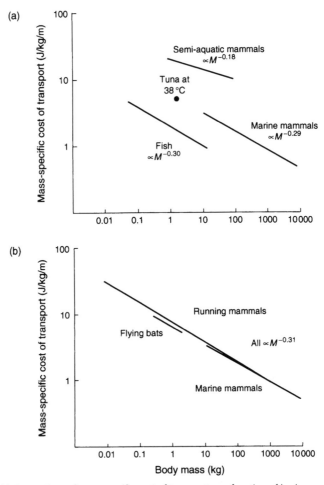

Fig. 9.15 (a) Comparison of mass-specific cost of transport as a function of body mass for semi-aquatic mammals, marine mammal swimming specialists and fish. (b) When the swimming transport costs of marine mammals are compared with those for mammalian runners and flyers (bats), similar scaling is observed for all groups. (Adapted from Williams (1999).)

do not incur the higher metabolic cost required to maintain an elevated body temperature in a cooler aquatic environment. Consistent with this, the resting metabolic rates of aquatic mammals, despite their greater insulation, are a factor of 1.7 to 2.4 greater than those predicted for terrestrial mammals of similar size (Williams 1999). Interestingly, tuna have a similar transport cost for their size as marine mammal swimming specialists. This probably reflects the higher metabolic cost associated with using warm internal red muscle for sustained active swimming.

9.8 Summary

Patterns of energy use during terrestrial locomotion exhibit remarkably similar relationships with speed and body size, despite enormous biological diversity in the number of limbs used to move, the type of skeleton used for support and the thermoregulatory strategy employed. The similarity of these patterns reflects the biochemical conservatism of the metabolic pathways within cells, as well as the organization and contractile properties of the skeletal muscles that power movement.

Differences in patterns of energy use are most extreme in terms of aerobic capacity and endurance. Differences also arise when comparing locomotion on land with that in water and air. Clearly, the trade-offs for greater or lower energy cost are balanced by the benefits that arise from the exploitation of the habitats that these locomotor modes afford. The tight link between energy use and locomotor performance reflects the central importance of how locomotor energy cost fits into an animal's overall energy budget. While minimizing energy use generally may be of considerable selective value, it is not the only aspect of performance that must be weighed in terms of understanding the locomotor design and capacity of an animal. Burst speed, maneuverability and strength, among others, are traits which must also be considered important to an animal's locomotor performance. Future work that explores how less regular intermittent patterns of activity affect energy use and performance will be important for furthering our understanding of patterns more characteristic of field activity. Finally, as exemplified by the comparison of swimming, flying and running mammals, it is also clear that evolutionary ancestry can have an important bearing on the energetics of locomotor performance.

10 | Neuromuscular control of movement

In previous chapters we have considered the organization and properties of the muscular and skeletal systems of animals which underlie their capacity for generating and supporting the forces necessary for locomotive movement. These are largely governed by the physical properties of the environments in which an animal moves. However, movement is only useful to an animal if it can be controlled. Movement often occurs in response to proximate sensory cues received from the environment. The ability to process sensory information and initiate a motor response is referred to as 'sensorimotor integration'. Motor responses may be fairly simple, involving local reflex pathways of feedback and control, or they may involve more complex longer-term responses which require the broader integration of higher centers within the nervous system. In both vertebrates and invertebrates the integration of auditory, visual, olfactory or other sensory stimuli is largely mediated by the brain to initiate and regulate motor responses appropriate to such stimuli. Although it is of central importance, sensory processing and motor integration at the level of the brain is quite complex and largely outside the scope of this book.

Instead, we shall focus on local levels of control involving fairly simple neural circuits. We shall explore how these local circuits can facilitate more decentralized coordination of locomotor rhythm and movement. This will involve first a consideration of the sensory 'transducers' located in the muscles, tendons and joints, and at the animal's body surface. These sensory elements monitor the animal's physical environment and the action of its muscles. The information received by these sensors is carried back to the animal's nervous system by **afferent** nerve fibers, providing feedback that is integrated at the level of the spinal cord of vertebrates and the sensorimotor ganglia of invertebrates. This results in an appropriate **efferent** output via motor nerves which innervate the muscles, controlling their action. Basic principles emerging from a consideration of reflex control of motor function at local levels can provide considerable insight into the fundamental requirements for coordinated and stable movement. In addition to local neuronal reflexes, it is likely that mechanical properties intrinsic to the animal's musculoskeletal system may provide an additional means for responding rapidly to changing demands and perturbations encountered during movement, simplifying the neural control task for maintaining balance and movement stability. However, such mechanisms are only now being considered and studied.

10.1   Sensory elements

Local control of muscle function and movement is mediated by various sensory components within both invertebrates and vertebrates. Three principal sensory components exist within vertebrates and two main sensory components exist within insects. Because of limitations of space and because insects represent the best studied invertebrate group, as well as representing species which exploit terrestrial, aerial and aquatic locomotive modes, our discussion here will focus largely on insects for comparison with vertebrate systems in order to highlight common principles of sensorimotor integration and function relevant to locomotive performance.

10.1.1   Vertebrate sensory organs

In vertebrates, two of the three main classes of sensory elements are located within the muscle itself, providing feedback of muscle length change (or muscle stretch) and muscle force. Muscle stretch is monitored by **muscle spindle organs** located in the belly of the muscle. Muscle force is monitored by **Golgi tendon organs**, which comprise free sensory nerve endings terminating at the muscle–tendon junction. Considered together, muscle spindles and Golgi organs provide the means for controlling motor output (force and length) in relation to movement and external forces imposed on the body. The third sensory modality involves a general class of **proprioceptive** feedback to the muscle, yielding tactile information via pressure and pain receptors located in the skin as well as mechanoreceptors located in the joints. These proprioceptive sensory elements will be discussed later in the context of the basic pathways of local neural circuits.

Muscle spindles

Muscle spindle organs (Fig. 10.1) comprise a set of sensory afferent nerves (Ia) and the special sets of modified **intrafusal** muscle fibers that they innervate. These intrafusal fibers are located in various locations within the belly of the

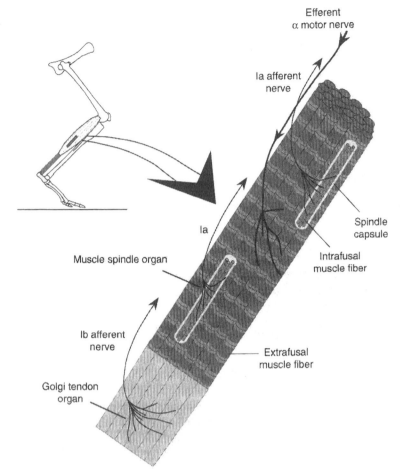

Fig. 10.1 Two classes of sensory elements exist within vertebrate muscle–tendon units: muscle spindles and Golgi tendon organs. Muscle spindles (commonly termed 'stretch receptors') transduce muscle length changes and are found in multiple locations throughout the muscle's belly. Golgi tendon organs (GTOs) transduce the force transmitted by the muscle and are found at the junction of the muscle's fibers with its tendon or aponeurosis. Muscle spindles are comprised of specialized non-force-generating 'intrafusal' fibers which relay changes of muscle length via Ia afferent nerves back to the spinal cord. GTOs relay force information via Ib afferent nerves. Of these two sensory receptors, the muscle spindles represent by far the most important means by which muscle force is regulated (via α motor nerve activation) in relation to muscle length.

muscle. They do not contribute to the force produced by a muscle and thus are distinct from the force-producing **extrafusal** fibers which comprise the bulk of a muscle. The Ia afferent nerve endings wrap in a helical fashion around the intrafusal fibers. (There are two types of intrafusal fibers—nuclear bag and nuclear chain—but differences in their response properties are not critical to our discussion here. Readers may wish to consult a standard physiology textbook for more details.) The Ia afferents possess stretch receptors located within the membranes of their dendritric receptive field. Thus strain of the muscle spindle intrafusal fibers stimulates the membrane stretch receptors of the Ia afferents, causing their activation. (In addition to the main Ia afferents, there are secondary type II afferent nerve fibers which also respond to stretch of the intrafusal fibers. However, we shall concentrate on the larger, more numerous and more important Ia afferents.) Although the intrafusal fibers do not contribute to force generation by the muscle, they retain the capacity to shorten, which is important to their ability to re-set their response to different ranges of length change. The intrafusal fibers are innervated by special motor nerves (gamma (γ) or 'fusimotor') which are distinct from the alpha (α) motor nerves. Recall that the α motor nerves are organized as motor units in relation to the pools of extrafusal fibers that they innervate. Many spindles (tens to hundreds) are found distributed throughout a muscle, allowing changes in muscle length to be monitored when different regions of a muscle are activated.

As we have discussed previously, when a muscle is activated to develop force its extrafusal fibers may remain isometric, shorten to produce a prescribed movement at a joint or be lengthened when resisting an external joint moment or force. If the external force resisting the muscle's action is of sufficient strength that it causes the muscle to be lengthened, the resulting stretch of the muscle will also stretch the intrafusal fibers in the muscle spindles. The Ia sensory nerve which innervates the intrafusal fibers responds to this stretch by increasing its firing rate (Fig. 10.2). The spindle organ also consists of intrafusal fibers that are sensitive to the rate of stretch. Consequently, muscle spindles provide feedback for both the magnitude and the rate of muscle stretch. The Ia afferent provides direct monosynaptic feedback to the motor nerves that innervate the muscle as well as synaptic input, via interneurons, to other muscle agonists and antagonists. This synaptic relay of the muscle's stretch due to an externally applied force occurs within the spinal cord (Fig. 10.3). As a result, the firing rates of the motor nerves that innervate the muscle (and possibly its agonist muscles) is increased, elevating the 'motor drive' to the muscle. This leads to additional recruitment of other motor units in the muscle, increasing the force that the muscle generates and enabling it to resist the stretch imposed by the applied load. A routine medical examination often used to test the integrity of this stretch response is to strike lightly the patellar tendon (in front of the knee). Such a test stretches and activates the spindles located in the knee extensor muscles (the quadriceps), normally resulting in knee extension and a forward kick of the leg.

Therefore muscle spindles provide an ongoing feedback of a muscle's length in relation to its activation as it resists external forces or moments. If a muscle remains isometric or shortens as it contracts, the firing rate of its spindle Ia

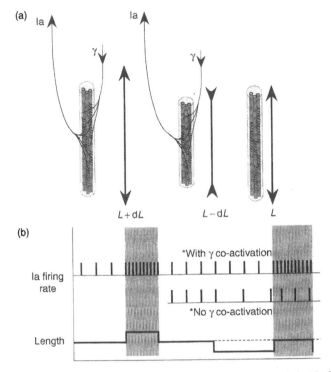

Fig. 10.2 The firing rate of Ia spindle afferents increases in response to stretch (+dL) of the intrafusal fibers (certain receptors within the spindle also respond to changes in the rate of stretch). The response of the spindle organ to changes in length can be adjusted by γ 'fusimotor' activation of the intrafusal fibers, causing them to shorten (−dL). This allows the spindle to respond to stretch when the muscle operates at shorter lengths. Without γ adjustment, stretch of the muscle from a shorter length would result in a weaker (slower firing rate) Ia response to stretch.

afferents will remain low or be proportionately reduced, leading to weaker feedback to the motor drive of the muscle (Fig. 10.2(a)). Modulation of spindle Ia afferent response to length change over the course of a contraction cycle can be achieved by simultaneous activation of the γ motor nerves, together with the recruited α motor nerves. This is often referred to as α–γ **motor co-activation** and is frequently observed for both voluntary and involuntary motor responses. This will cause the intrafusal fibers to shorten along with the shortening of the whole muscle (Fig. 10.2(b)). Shortening of the intrafusal fibers means that spindle Ia afferents will shift their firing rate response to operate at shorter muscle lengths. By relaxing the intrafusal fibers through inhibition of the γ motor nerves, the spindle response to length change will shift in an opposite fashion, allowing more effective sensory feedback of the spindle Ia afferents when a muscle operates at longer lengths.

Segmental Spinal Cord

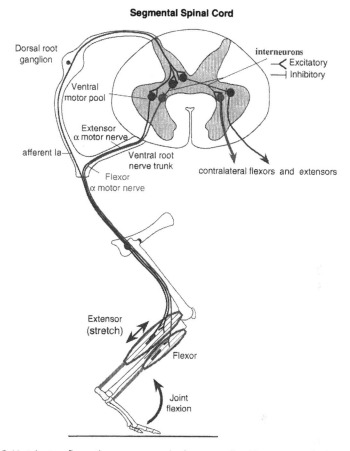

Fig. 10.3 Vertebrate reflex pathways are organized segmentally with respect to the limb and spinal column. Afferent nerve fibers carrying sensory information from muscle spindles, Golgi tendon organs and proprioceptors have their cell bodies in the dorsal root ganglion and transmit their information to interneurons within the spinal cord. These interneurons relay sensory information to other levels (and limbs) within the spinal cord, as well as to opposing muscles within the same limb. This is mediated by synapses with the dendrites of α motor neurons which innervate the limb muscles that control movement and body support. Spindle Ia afferents synapse directly back onto α motor neurons of the same (hymonomous) muscle, forming a monosynaptic pathway which facilitates a rapid motor response to stretch of the muscle. Interneurons mediate reciprocal inhibition (or activation) of opposing sets of muscles within and between limbs.

The ability to modulate spindle Ia afferent response is important for enabling muscles to control movements over differing ranges of length change, depending on the motor tasks involved. When one realizes that many muscles often undergo time-varying patterns of stretch, force development and shortening over the course of a single contraction cycle (see Figs 2.5, 3.13 and 5.10), the role of spindles for providing length feedback is key to a vertebrate muscle's ability to control length and position, as well as the manner in which it develops force and does mechanical work. Because spindle Ia feedback involves a monosynaptic pathway, this minimizes the time delay between sensing stretch within the muscle and the recruitment of increased muscle force. This is probably important for the control of balance during a limb's support phase. In general, muscles which control precise movements, such as those controlling the fingers, have a higher density of spindles than those which control grosser movements of the body.

Golgi tendon organs

Golgi tendon organs consist of Ib sensory afferents which have free nerve endings innervating the collagenous connective tissue of the muscle aponeurosis and tendon (Fig. 10.1). Compared with muscle spindles, muscle Golgi tendon organs are much simpler in their organization and in how they function. Golgi organs are most common and have been studied in pinnate muscles which attach via tendons to the skeleton. Golgi organs monitor muscle force by sensing the strain developed in the muscle's aponeurosis or tendon in response to the force the muscle develops. Relatively few, if any, Golgi organs are found in parallel-fibered muscles which have little tendinous or aponeurotic insertion on the skeleton. Consequently, their role in monitoring force in parallel-fibered muscles is minimal.

Similar to the spindle Ia afferents, the Golgi Ib afferents sense force by means of membrane stretch receptors located within their dendritric field. Strain within the muscle–tendon junction stimulates the membrane stretch receptors, leading to depolarization and activation of the Ib sensory nerve in response to increasing force. The firing rate of the Golgi Ib afferents increases proportional to the force that a muscle develops. In general, Golgi tendon organs are believed to exert an *inhibitory* feedback to a muscle and its synergists, limiting the force that the muscles develop, which could be damaging if excessive. Thus, in principle, motor drive to a vertebrate muscle involves dual control: excitatory via the spindles and inhibitory via the Golgi tendon organs. However, available evidence indicates that Golgi organ inhibition is typically not very strong over much of a muscle's force range. Hence, spindles are generally believed to exert the main control of a muscle's motor drive for most behaviors.

10.1.2 Insect sensory organs

Insects possess two classes of sensory receptors, commonly referred to as **exteroreceptors** and **proprioreceptors**. Exteroreceptors reside on the surface of the cuticle as hair-like projections of different types which occur in varying densities and distributions on the animal's body. Like the proprioceptive skin receptors of vertebrates, the majority of insect exteroreceptors sense tactile information which provides a mechanism for detecting when a limb has contacted the ground or when it has bumped into an object during movement (others sense chemical stimuli or air movements). These receptors can also often provide directional sensitivity in response to displacements of the hair in differing directions (Burrows 1996).

The proprioceptive elements of insects generally act as strain gauges (i.e. mechanoreceptors) which respond to the forces and pressures exerted by muscles internal to the insect's exoskeleton and those resulting from movement at the joints. Proprioceptive elements located within the cuticle are generally referred to as 'campaniform sensilla' (singular, sensillum), because of their bell-like appearance (Fig 10.4). Proprioreceptors located within joints are represented by chordotonal organs and other types of joint receptors. Proprioreceptors are less numerous than exteroreceptors and are more densely distributed in areas near the joints where the muscles and apodemes insert onto the cuticle. The campaniform sensilla are characterized by canals extending through the cuticle, through which dendrites of sensory neurons pass to attach to a thin membrane at the surface of the receptor (Fig. 10.4). Thus localized deformations of the cuticle are sensed by the strains imposed on the sensory nerve endings of the campaniform sensillum.

In the insect leg, these strains are produced by the forces transmitted by the underlying muscles and apodemes which stabilize and move the exoskeleton. In the wing of flying insects, the campaniform sensilla represent a network of pressure receptors which provide sensory feedback of the wing's deformation in response to aerodynamic loads. Similarly, sensilla also provide sensory feedback to pressures developed in response to joint movement. Unlike vertebrates, there are few known examples of insect muscles, or other invertebrate muscles, which possess a length-sensing element comparable to the spindle organs of vertebrate muscles (one exception is the abdominal musculature of crustaceans, such as lobster and crayfish, which has stretch receptors that are important to control of the alternating dorsoventral beating movements of the tail associated with the animal's escape response (Hoyle 1983). Hence, whereas vertebrate muscles are largely under length control via muscle spindles, the insect muscles are largely under force or force-based motion control derived from the feedback provided by campaniform strain receptors.

10.2   Sensorimotor integration via local reflex pathways

Control of muscle function by proprioceptive (tactile), force and length reflex pathways provides local feedback to the motor neurons that are activated and

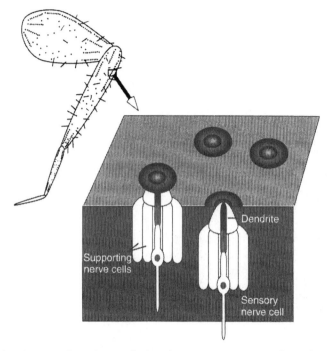

Fig. 10.4 Insect sensory elements comprise two classes: exteroreceptors and proprio-receptors, represented here by hair receptors and campaniform sensilla, which are distributed over.

responsible for a particular motor action. Local reflex control is advantageous because it facilitates rapid motor responses to sensory stimuli and does not require higher-level signal processing by the central nervous system. Although critical for responding to more complex stimuli and for organizing and initiating more involved and prolonged motor behavior, central nervous system processing takes longer to accomplish and requires greater attention to the motor task. This would not be an effective means for dealing with the rapid and momentary disturbances that an animal often encounters when moving through its environment. Such disturbances are much better dealt with at a 'less conscious' and more local level of control. Thus, local reflex pathways are an important component of the distributed nature of neuromuscular control, which seems to be a general design principle of most motor systems. We shall begin by discussing vertebrate reflex pathways, as these are the best studied and highlight principles of neuromotor organization and control which are probably common to a diversity of motor systems.

10.2.1   Vertebrate reflex pathways

Sensory information from proprioceptive elements, muscle spindles and Golgi tendon organs all passes through the paired segmental dorsal root ganglia which lie on either side of the vertebrate spinal cord (where their cell bodies reside) and enters the spinal cord to synapse onto interneurons or, in the unique case of the spindle Ia afferents, directly onto motor nerves innervating the same ('hymonymous') muscle (Fig. 10.3). Most sensory feedback is transmitted via one or more interneurons before being sent to motor neurons innervating the muscles within the limb. Interneurons also provide sensory feedback to motor neurons which innervate muscles of the opposing (contralateral) limb. In this way, sensory information from one limb or side of the body influences, and can be coordinated with, the motor activity of the opposing limb or side of the body. In vertebrates, the motor neurons are concentrated within motor pools located within the ventral region of the spinal cord. These motor pools are segmentally arranged within the spinal cord in association with the location on the body axis where the limb is found and where the muscles of the limb are arranged (e.g. muscles which protract the limb are often associated with more cranial levels of motor innervation, whereas muscles which retract the limb are often associated with more caudal levels).

With foot contact on the ground, pressure receptors within the base of the foot provide proprioceptive feedback via the dorsal root ganglia. This feedback, together with increased stretch activation of the spindle organs of certain limb extensor muscles, may provide excitatory feedback to increase the motor activation of these muscles (Figs 10.2 and 10.3). This increase in activation involves an increased firing frequency of motor units that are already active, as well as the recruitment of additional motor units to increase the total force output of the recruited muscles. Thus, as a limb progressively experiences increasing load during its support phase, proprioceptive and spindle stretch receptor feedback act to stimulate more forceful limb extension. In contrast, if a sharp pain is felt, rather than stimulating a more forceful activation of the extensor muscles (during limb support), pain receptors initiate a 'withdrawal reflex'. This involves reduced excitation, or even complete inhibition, of limb extensors and activation of limb flexors in order to shift weight support away from the limb contacting the painful stimulus. To maintain balance, such a withdrawal reflex requires the coordinated activation of extensor muscles of the opposing limb(s) to allow for the shift in weight support.

While useful and reasonably accurate, it should be recognized that these model descriptions for increased feedback and motor stimulation in response to load are probably oversimplified. In the actual movements of animals, the regulation of motor recruitment involves a more complex interaction of reflex feedback from various sensors within a limb (and among different muscles).

In combination with descending control by the central nervous system, much of the phasic activity of limb muscles, i.e. their oscillating flexor–extensor activation, is also derived from local levels within the spinal cord (see section 10.5). For example, the activation of certain limb extensors may show biphasic stimulation patterns, allowing the limb to flex momentarily during the stance phase of the stride. As we discussed in Chapter 3, this is important for reducing the potential energy work of the body's center of mass as it passes over the supporting limb when an animal walks. Modulation of motor activation is also probably linked to the temporal pattern of force development of a muscle and how this influences whether its role is to produce mechanical energy by shortening, absorb energy by lengthening, or contract isometrically to facilitate spring-like function of the tendons and the limb as a whole. Hence the actual integration of proprioceptive, spindle and Golgi organ feedback with centrally coordinated patterns of motor drive is more complicated than the simple schemes presented above. Nevertheless, they provide the basis for understanding how higher-level coordinated control of locomotive function is achieved.

Muscle stiffness: a control objective of motor function?

An attractive aspect of the reflex feedback provided by vertebrate spindles is that this may provide a means for regulating muscle stiffness ($=\Delta F/\Delta L$). By adjusting changes in muscle force (neuromotor drive) to changes in muscle length, the nervous system's task of controlling muscle function may be made easier by the simple objective of achieving and maintaining a constant muscle stiffness (Houk 1979). Such a model requires that any increase in force that a muscle develops is matched to a given increase in muscle stretch and feedback by the muscle spindle Ia afferents. This would maintain a linear relationship between force and length (as from point A to point B in Fig. 10.5). By maintaining a constant muscle stiffness, mechanical perturbations of joint and limb displacements, which might otherwise be destabilizing, can be resisted in a proportional fashion. This is an attractive model because it involves a linear response of motor activation which is far simpler than non-linear control strategies. However, evidence for this model remains limited and needs further testing. Nevertheless, it provides an attractive rationale for how muscle force and length may be regulated to resist unexpected perturbations of musculoskeletal force transmission.

10.2.2 Insect reflex pathways

Sensory force information from surface hairs, campaniform sensilla and other insect mechanosensory elements provides feedback via reflex pathways which mirror the general organization described for vertebrates. Whereas sensory

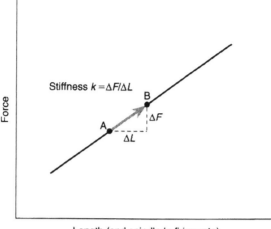

Fig. 10.5 Control of muscle stiffness in vertebrates is mediated via the muscle spindles. Linear responses of spindle Ia discharge rate in response to length changes produced by imposed increases in muscle force is hypothesized to provide a means for maintaining muscle stiffness constant (linear slope). This has been suggested to be a means for simplifying the control of a muscle's contractile function.

feedback from hair exteroreceptors is nearly always indirect to the motor neurons, feedback by mechanosensory neurons is generally direct (monosynaptic), similar to the spindle Ia afferents of vertebrates. Insects (and other arthropods) possess a ventral nerve cord which consists of segmental ganglia that provide local integration of opposing limb (and wing) function (Fig. 10.6). As in vertebrates, interneurons link motor neuron pools to provide coordination of contralateral limb movements. In addition, intersegmental interneurons running between segmental ganglia underlie control of interlimb coordination for ipsilateral limbs (front, middle and rear of insects).

In contrast with vertebrates, the number of sensory afferents in the limbs of insects greatly outnumbers the number of motor neurons that control their muscles. Consequently, there is a large-scale convergence of the many thousands of sensory afferents in the connections made with the much smaller number (few hundreds) of motor neurons that control the limb muscles. This reflects the fundamentally different motor unit organization of vertebrates compared with that found in arthropods and other invertebrates (see below). This convergence largely occurs via a special class of 'non-spiking' interneurons (in addition to spiking interneurons) located within and between the ganglia (Burrows 1996). As their name suggests, non-spiking interneurons do not fire action potentials when depolarized (in contrast with the 'all-or-none' properties of vertebrate neurons). Instead, they exert graded voltage

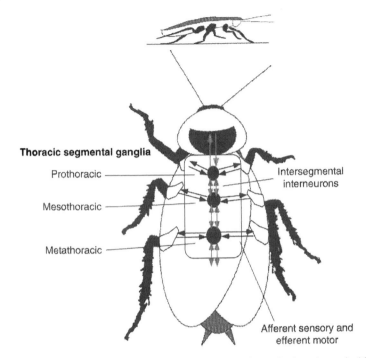

Fig. 10.6 Insect (cockroach) reflex pathways parallel the general organization observed within vertebrates (see Fig. 10.3). Three ventral motor ganglia reside within the thorax of insects. Each motor ganglion controls the relative timing of flexor and extensor activity within and between its associated paired set of limbs. Afferent sensory information from exteroreceptors and proprioreceptors (found within the joints and the muscle's apodemes) is relayed back to the motor ganglia where it converges on interneurons that control the output to motor nerves supplying the limb muscles (see Fig. 10.7). Interneurons also relay local sensory information to the other motor ganglia to coordinate the relative timing of muscle activation between sets of limbs. Visual, olfactory and tactile (via antennae) sensory information is also relayed from higher centers within the head via interneurons to the thoracic motor ganglia to control the overall motor response of the animal.

effects on the motor neurons and other interneurons which they innervate. Consequently, non-spiking interneurons can be viewed as cellular integrators which process the wide-ranging input (from the same and other limbs) that converge onto them from a large number of sensory afferents. As such, they serve to regulate the output to the various motor pools that control and drive the muscles of the limbs. Evidence suggests that particular non-spiking interneurons may control specific motions of a leg.

In almost all cases, the input from proprioceptive and exteroceptive sense organs is excitatory to the interneurons and motor neurons of the segmental ganglia (A and B in Fig. 10.7(a)). Activation of mechanosensory neurons derived from campaniform sensilla or chordotonal organs which sense joint motion or force associated with extensor muscle activity may provide positive excitatory feedback to the extensor muscles of the limb, increasing their resistance to joint flexion and providing greater weight support. Their feedback to motor neurons innervating limb flexors is inhibitory, mediated via the inhibitory synapses of interneurons within the ganglion controlling the limb (C in Fig. 10.7(a)). In addition, inhibitory feedback via interneurons within a ganglion acts on the appropriate muscles of the opposing limb to reduce their role in weight support or to trigger their transition to a swing phase of the locomotor cycle (D in Fig. 10.7(a)). Similar to vertebrates, the inhibitory action of interneurons thus provides reciprocal inhibition to control the appropriate

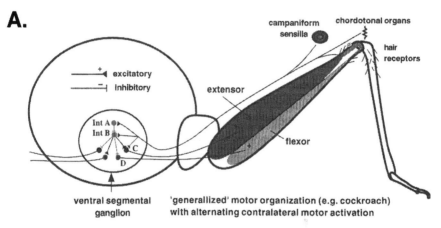

A.

Fig. 10.7 A) Generalized reflex sensory-motor pathways to the hind leg of an insect. A key difference from vertebrates is that sensory afferents converge on interneurons which integrate and distribute their output to motor nerves to control specific motor behaviors in response to stimuli. Few afferents synapsse directly with motor nerves (C, extensor & D, flexor). Sensory afferent nerves from hair receptors, campaniform sensilla, chordotonal organs, and other receptors converge on spiking (Int A) and non-spiking interneurons (Int B) in the ventral segmental ganglion. Spiking interneurons also inhibit non-spiking interneurons. In walking and running insects, such as cockroaches, interneuron stimulation to contralateral extensor muscles is largely inhibitory reflecting the out-of-phase nature of limb movement patterns similar to that observed in walking and running vertebrates. B) In locusts and grasshoppers the enlarged hind leg is used for jumping as well as for slower movements, in which the limbs must move synchronously. The extensor tibiae muscle is innervated by two motor nerves, SETi (slow) and FETi (fast), which receive largely excitatory stimulation via populations of metathoracic interneurons (Int A and Int B) from sensory receptors in the limb. As discussed in Chapter 7 (section 7.3.1), jumping in these insects involves initial activation of the flexor to flex and lock the hind leg at the knee joint, subsequent co-activation of the extensor and flexor, allowing the extensor tibiae to develop isometric tension and store elastic energy in the cuticle, followed by a trigger release via common inhibition (CI) of the flexor motor neurons by as yet unidentified interneurons (Heitler & Burrows, 1977). The synchronous movements of the hind legs of locusts, grasshoppers and other jumping insects is therefore mediated by a common inhibition of the flexors of both limbs simultaneously to release the catch of the hind limbs for the jump.

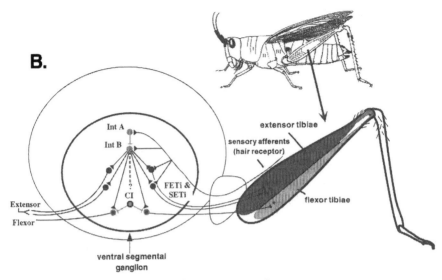

Fig. 10.7 continued

action of the antagonistic muscles within a limb, as well as the actions of muscles in the other limbs. Finally, insects also display a withdrawal reflex response similar to that of vertebrates, in which the limb may be retracted away when a noxious stimulus is applied to a particular appendage.

Local reflex pathways which control the flexors and extensors of a given limb are linked via the network of interneurons within and between each ganglion to control the relative timing of muscle activation among the various limbs associated with a given gait. Since the relative phase and duration of arthropod limb movement changes with gait, as it does in vertebrates, the interacting influences of local reflex pathways within a limb must be coordinated with and clearly influence the control of motor activation in opposing and ipsilateral limbs. Although initiation of a particular limb-movement pattern may occur in response to sensory stimuli received by higher centers within the nervous system, maintenance or modulation of a particular pattern in response to external perturbations or local stimuli is also probably a major component of an effective decentralized motor control system. Furthermore, while the finely graded control necessary to maintain balance when an insect is moving rapidly almost certainly involves a complex integration of sensory and motor signals via local reflex pathways, recent work in running cockroaches suggests that intrinsic biomechanical properties of animal limbs may help provide for stabilizing control in response to disturbances of an animal's balance (Jindrich and Full 1999), obviating or simplifying the need for more centralized nervous system control.

Table 10.1   Motor innervation features of invertebrate versus vertebrate skeletal muscles

Feature	Invertebrate	Vertebrate
Synaptic input	Excitatory, inhibitory and modulatory	Excitatory
Neurotransmitter	L-glutamate[a], acetylcholine, GABA and octopamine	Acetylcholine
Motor nerve terminals	Multiple and distributed	Single or few, and localized
Muscle fiber innervation	Polyneuronal	Single neuron[b]
Activation	Graded and twitch	All-or-none, twitch
Force modulation	Graded and frequency	Motor unit recruitment

[a] L-glutamate is found in arthropods, but acetylcholine is the excitatory neurotransmitter in annelids, molluscs and echinoderms.

[b] A few slow fibers of some fish and amphibian muscles, and eye muscles of mammals, receive polyneuronal innervation.

10.3   Muscle recruitment in relation to functional demand: force, speed and endurance

We now turn our attention to the neuromuscular organization of vertebrate and invertebrate muscles, which underlies the means by which motor recruitment is regulated for meeting the functional demands of changes in force output, speed and endurance. Considerable differences exist between vertebrates and invertebrates regarding the manner in which the nervous system innervates and regulates motor recruitment (Table 10.1). This is believed to reflect, at least in part, the constraints of size imposed on the organization of the nervous system in invertebrates, which are in general much smaller than vertebrates.

10.3.1   Vertebrate motor recruitment

Motor innervation of vertebrate muscles is exclusively excitatory in nature and occurs via the neurotransmitter acetylcholine (Table 10.1). In almost all cases, innervation of vertebrate muscle fibers is via a single motor nerve which makes a connection with the fibers via a single local synaptic endplate junction. In contrast with many invertebrate motor junctions, activation of vertebrate muscle fibers is all-or-none (i.e. 'twitch'), in which the depolarization of a fiber results in a rapid spread from the synaptic endplate along the fiber's length. This depends on the presence of rapidly conducting voltage-dependent sodium channels in vertebrate muscle fibers which are not found in invertebrate muscle fibers.

As discussed in Chapter 2, vertebrate muscles are comprised of subpopulations of twitch fibers which can be classified into three principal types. In most muscles, and almost exclusively in mammals and birds, a single motor nerve innervates a distinct population of fibers of a single type forming a motor unit (Fig. 2.7). Consequently, changes in force, speed and endurance are mediated by the manner in which the nervous system recruits different motor units within the muscle. The number and size of motor units varies among different muscles, ranging from as few as 10 or 20 to several hundred motor units per muscle. In general, smaller muscles consist of smaller motor units (i.e. relatively fewer numbers of fibers innervated per motor neuron). This allows for more fine-grained control of force and speed via recruitment of the muscle's motor unit pool. For example, the small muscles which control the fingers are comprised of hundreds of small motor units. This is critical to the ability to grasp and manipulate objects, as well as for performing motor tasks such as playing a musical instrument or typing on a keyboard. Whether or not scale effects on motor unit organization exist for muscles from animals of very different sizes is not very well known. Partly, this reflects the fairly daunting task of identifying and counting individual motor units within a muscle. Measurements of motor unit size (number of fibers per motor unit) relative to differences in muscle size may give some indication of scale effects on motor unit organization. For example, the tibialis anterior muscle of a rat is approximately 10 times smaller than that of a cat, but the average motor unit size within the muscle of these two species varies by less than twofold. This suggests that the number of motor units may decrease with body and muscle size (in this example, a fivefold reduction) when comparing across species.

The organization and properties of mammalian motor units have been the best studied and will be the focus of our discussion here. However, it should be noted that, occasionally, individual muscle fibers of other vertebrates (mainly fish and amphibians) may be innervated by more than one motor nerve, as in invertebrates. In such cases, polyneuronal innervation provides a means by which the activation of a muscle fiber may be influenced jointly by the neural recruitment of more than one motor nerve.

Orderly recruitment: the 'size principle'

Because of their uniform fiber-type composition, vertebrate motor units exhibit characteristic properties which reflect those of their fibers (see Table 2.1). Motor units comprised of slow-oxidative (SO) fibers are generally small, having a small-diameter motor nerve axon which innervates a small number of fibers with relatively small cross-sectional areas (Table 10.2) (see also Fig. 2.9). In contrast, fast-glycolytic (FG) units are large, composed of a larger-diameter motor axon which innervates a large number of larger-diameter fibers. Consistent with their metabolic and contractile properties, fast-oxidative-glycolytic (FOG) motor units possess intermediate organization. These differences in the relative sizes of the motor nerves and the total fiber area represented by each motor unit influence the excitability and the level of force output that can be recruited

Table 10.2 Vertebrate skeletal muscle motor unit features based on fiber type

Feature	Slow-oxidative (SO) (Type I)	Fast-oxidative glycolytic (FOG) (Type IIa)	Fast-glycolytic (FG) (Type IIb)
Innervation ratio[a] (motor unit size)	Low	Intermediate	High
Motor nerve axon diameter and cell body size	Small	Intermediate	Large
Stimulation frequency	Low	Intermediate	High
Excitability	High	Intermediate	Low
Inhibitability	Low	Intermediate	High
Contractile			
Shortening speed	Slow	Moderate to fast	Fast
Fatigue rate	Low	Intermediate	Fast

[a]number of muscle fibers per motor nerve.

via activation of a muscle's motor neuron pool in the spinal cord. At low levels of activation, the most excitable low-diameter motor axons and smallest motor units are the first to be recruited. As activation intensity increases (increased firing rate and hence greater summed synaptic excitatory input to the motorneuron) there is an orderly progression of recruitment from small to larger motor units as a greater number of the motor units within a muscle's spinal cord pool are recruited. This means that the smallest, most oxidative and high endurance (SO) fibers within the muscles of the limb are recruited first, under conditions which typically require low levels of force output to be exerted over longer periods of time (e.g. when moving slowly, shifting balance or maintaining posture). When more rapid and forceful movements are required, recruitment shifts to larger and faster motor units which possess less oxidative and more glycolytic capacities (FOG and FG). This orderly recruitment of muscle fiber types from small (slow) to large (fast) was first recognized by Henneman and coworkers (Henneman 1957; Henneman *et al.* 1965). It is now widely recognized as the 'size principle' of motor unit recruitment. Although certain instances have been observed in which FG units are recruited before FOG or SO units, orderly recruitment according to motor neuron and unit size appears to hold quite generally for a broad range of motor tasks in diverse vertebrate groups. It also explains well the shifts in motor unit and agonist muscle activation that are observed as animals increase their speed and change gait during locomotion.

Motor unit distribution

Motor units have varying distributions within different vertebrate muscles. The region of a muscle over which the fibers of a given motor unit are distributed constitutes a motor unit's **territory**. In general, motor unit territories are often distributed over a third of the muscle's total cross-section. Consequently, in most cases motor unit territories greatly overlap with one another. However, in some muscles, such as the lizard iliofibularis, their distribution may be quite distinct, with one (central) region being composed of SO units and another (outer) region exclusively composed of FG units. Because motor units greatly overlap in most muscles, their organization is typically heterogeneously distributed throughout the muscle's cross-section (see Figs. 2.7(a) and 2.9). This means that as recruitment in a muscle shifts from slower to faster units, the force that individual units produce is probably summed in a fairly uniform fashion throughout the whole of the muscle.

For muscles with less heterogeneous fiber-type distributions, in which motor units with different properties are compartmentalized into different regions, differential recruitment within the muscle can be expected to result in a more regional summation of force and length change within the muscle as a whole. However, the extent to which this occurs will depend on the degree to which force generated by active muscle fibers is transmitted to surrounding connective tissue components of the fibers. This will favor a more generalized transmission of force within the muscle, even if its motor unit organization is compartmentalized into different regions.

Different muscles within an agonist group can also often have differing motor unit compositions. An extreme and classic example of this is the triceps surae, or ankle extensor muscles, of the cat. The triceps surae is a group of leg muscles comprised of the medial and lateral heads of the gastrocnemius which, together with the soleus, transmit their force via the Achilles tendon to extend the ankle joint. Whereas the medial and lateral gastrocnemius muscles are comprised mainly of FOG and FG fibers and a few SO fibers, the soleus is exclusively composed of SO fibers. Such differences in motor unit organization among functional agonists provides a means by which motor recruitment can be geared to changing demands of locomotive speed and force. In the cat (and many other mammals), the soleus is recruited for postural control and is the main ankle extensor muscle that is activated during walking. When cats increase their speed by changing gait to a trot or gallop, or when they jump, the gastrocnemius muscles are recruited to provide more rapid and forceful ankle extension. Shifts in motor recruitment, both within and among muscles, also affect the endurance capacity of the animal associated with a given level of physical activity. As faster contracting but more fatiguable motor units are recruited, the endurance capacity of the animal is reduced.

10.3.2 Invertebrate motor recruitment

In contrast with vertebrate motor innervation, which is exclusively excitatory in nature, motor innervation in insects and other invertebrates involves both excitatory and inhibitory synaptic input (Table 10.1). Excitatory motor junctions in arthropods utilize L-glutamate as the neurotransmitter, whereas inhibitory motor neurotransmision is via γ-aminobutyric acid (GABA). Many arthropod muscles also receive a third modulatory synaptic input via octopaminergic neurons, which were originally thought to enhance the excitatory input to limb muscles through their secretion of otopamine locally on motor junctions, but are now recognized to contribute more complex neuromodulatory input to motor nerves and probably interneurons as well (Burrows and Pfluger 1995). In annelids, molluscs and echinoderms, acetylcholine serves as the excitatory neurotransmitter, similar to vertebrate motor synapses.

Synaptic input from the motor nerves typically involves multiple terminals which have many branches that are widely distributed over the fiber's surface. In distinct contrast with vertebrates, the neural activation of many invertebrate muscle fibers is graded locally within the muscle fiber, as a result of the summed excitatory (and inhibitory) potentials which the various nerve terminals transmit locally to the fiber. Many insect motor nerves exhibit 'spiking' properties, but in most instances these reflect local depolarization of the endplate junction and not an all-or-none activation of the whole muscle fiber. Therefore the many-branched multiple terminal endings of the motor nerves are required for contraction of the whole fiber. Consequently, in contrast with vertebrate motor recruitment, motor recruitment in many invertebrate muscles is largely modulated by differences in the phase and frequency of motorneuron stimulation of a muscle, which influences its level of depolarization and yields a graded response in terms of force output. This is correlated with the fact that insect muscles are commonly innervated by only one or two (and never more than nine) motor nerves (Hoyle 1983).

Hence, control of muscle force and speed must be achieved within the context of how only a few motor nerves activate the muscle. This reflects a much simpler organization for motor recruitment in comparison with the graded recruitment of many hundreds of motor units which often underlies the control of vertebrate muscles comprised of different muscle fiber types (Table 2.1). It also provides a more economical organization, requiring far fewer motor nerves to control the contractile function of a muscle. Consequently, invertebrate motor recruitment depends on the graded level of activation of much larger fractions of the muscle as a whole, rather than the summed recruitment of many individual motor units, as in vertebrates. This probably reflects the important constraint of size faced by small animals like insects and many other invertebrates.

Invertebrate muscle fiber types

Like vertebrates, different types of fibers are found within the locomotor muscles of insects and other invertebrates. However, the limited number of

Table 10.3 Invertebrate skeletal muscle fiber types (based largely on arthropods)

Feature	Slow	Intermediate	Fast
Myosin ATPase	Low	Intermediate	High
Metabolic enzymes	Aerobic	Mixed aerobic and glycolytic	Glycolytic[a]
Mitochondria	Numerous	Moderate/variable	Few
Sarcoplasmic reticulum	Sparse	Moderate	Extensive
Contraction speed	Slow	Moderate	Fast
Fatigue rate	Low	Moderate	Rapid
Motor nerve			
Motor nerve axon diameter and cell body size	Small	Intermediate	Large
Excitatory synaptic potential	Small	Medium	Large
Excitability	High	Intermediate	Low

[a] Except insect flight muscle.

muscles that have been studied and the varied properties of their fibers makes it difficult to classify them other than very generally and largely in terms of studies of arthropods (Table 10.3). Similar to vertebrates, three general classes of fibers are distinguished: slow, intermediate and fast. However, it is important to distinguish and note that the 'slow' fibers of invertebrates are often much slower contracting and more resistant to fatigue than the slow 'twitch' fibers of vertebrates. As a consequence, they are often referred to as being 'tonic'. In addition, as with vertebrate muscle fibers, differences in contraction rate and susceptibility to fatigue (fast fibers being most readily fatigued) correlate with differences in their enzyme characteristics and their cell architecture (extent of sarcoplasmic reticulum and number of mitochondria). As was noted in Chapter 2, a key difference between invertebrate and vertebrate muscle fibers is their sarcomere length. Whereas sarcomere length is fairly uniform among vertebrate skeletal muscles (2.2–3.0 μm), it varies considerably among invertebrates; ranging from 2.0 to 13 μm in arthropods and up to 40 μm in annelids. In general, muscles with longer sarcomere lengths contract more slowly than those with shorter sarcomere lengths.

Example of neuromotor organization in the locust and cockroach.

In terms of neuromotor recruitment and control, locusts and cockroaches are two of the best-studied invertebrate species. Each of these insects possesses paired sets of prothoracic, mesothoracic and metathoracic ganglia that control the movements of their fore, middle and hind legs (Fig. 10.6). The extensor tibiae muscle, which is located within the femur and extends the tibia (Fig. 10.7B), or distal portion of each leg, is the best-studied muscle. In both species the extensor tibiae of each leg is innervated by a pair of excitatory nerves, a slow excitatory nerve (SETi) and a fast excitatory nerve (FETi), which emanate from the motor ganglia of each limb (Hoyle, 1983). In addition, the muscles of all three pairs of limbs are innervated by a octopaminergic modulatory interneurons. Activation of the motor nerves controlling the muscles of each leg is largely integrated via afferent input to both spiking (Int A, Fig. 10.7) and non-spiking (Int B, Fig. 10.7) populations of interneurons within each ganglion (Burrows, 1989). This is distinct from vertebrates in which spindle Ia afferents synapse directly on to homonymous motor neurons. Spiking interneurons also synapse on to intersegmental interneurons (Fig. 10.6) to integrate movements among sets of legs. In slow movements and during walking, the slow extensor nerve is mainly activated; however, some more limited fast extensor activation may also be observed. As speed increases, fast extensor activation increases. Although the roles of modulatory interneurons are still unclear, they are believed to serve a similar role as adrenaline (or epinephrine) in vertebrates, to facilitate a rapid escape response by the animal.

In locusts and grasshoppers, which have evolved a greatly enlarged metathoracic or hind limb for jumping, the fast FETi nerve is only activated to power rapid contraction of the extensor tibiae when the animals jump, whereas the SETi controls slow movements of the animal's hind leg, as well as being activated during jumping. Sensory input mainly from hair receptors on the leg or abdomen initially results in flexion of the knee by activation of the flexor tibiae, which is followed by co-activation of the extensor tibiae to both hind limbs (Fig. 10.7; Heitler & Burrows, 1977). Owing to the greater mechanical advantage of the flexor tibiae (section 7.3.1) this keeps the knee locked into position allowing the extensor tibiae to develop maximal isometric tension and store elastic energy in the cuticle at the knee. This is followed by a 'triggered' release of the catch mechanism by simultaneous inhibition of the flexor tibiae muscles to both hind limbs. This is mediated by an unidentified set of common inhibitory interneurons (CI). The advantage of having a separate trigger release via inhibition of the flexor muscle, rather than control via a change in excitation to the motor nerves, is that it assures precise timing of the release of elastic strain energy from both hind limbs and a more stable jump.

Invertebrate muscle activation patterns in relation to speed

When insects move more quickly the frequency of motoneuron bursts controlling different limb muscles and the firing rate within each burst generally both increase, similar to the pattern observed in vertebrates (Fig. 10.8). This causes the muscle to generate more force at a higher rate, leading to more forceful and

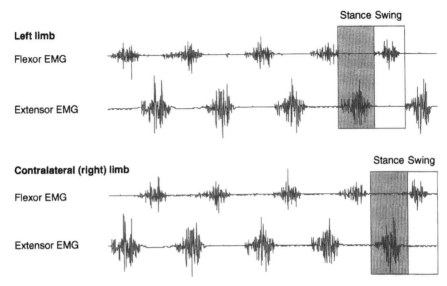

Fig. 10.8 Representation of flexor and extensor EMGs of the right and left limbs of a walking or running animal showing the out-of-phase flexor–extensor timing which occurs within a limb and between limbs. This is achieved via reciprocal inhibition by interneurons located within the spinal cord of vertebrates or thoracic motor ganglia of insects. Activation of extensor motor nerves during the stance phase inhibits flexor motor nerves supplying muscles of the same limb (antagonistic inhibition), but stimulates the flexor motor nerves and inhibits the extensor motor nerves of the opposite limb.

rapid limb movements. There is little evidence that insects and other invertebrates have much ability to regulate their endurance capacity for activity at differing intensities during locomotive movement (other than, perhaps, by strategies of intermittent locomotion (see Chapter 8, section 8.8)). This is largely because their ability to recruit different populations of muscle fibers within individual muscles which have different contractile properties is limited by their simple motor unit organization. However, in studies of the swimmerets of lobsters, which are innervated by three shared excitatory nerves that produce motor synaptic potentials of different sizes (small, medium and large), there is evidence of recruitment to control slow, moderate and fast swimming speeds by progressive activation of the three excitatory motor nerves (Hoyle 1983). The fastest excitatory nerve with the largest synaptic motor potential is also the one that fatigues most quickly, whereas the slow excitatory nerve with the smallest motor potential is resistant to fatigue. Finally, similar to a general pattern observed in mammalian motor recruitment (see below), the order of recruitment of the excitatory nerves appears to follow a progressive sequence from slow to medium to fast. This is also similar to the pattern observed in ghost crabs, in which only the slow excitatory nerves to the principal limb muscles are activated during walking, whereas the larger fast excitatory nerves are recruited when the crab runs (Burrows and Hoyle 1973).

10.4   Reciprocal inhibition: a basic feature of sensorimotor neural circuits

Two features of locomotive function underlie the more general organization of the sensory and motor systems of the body axis and limbs of both vertebrates and insects (as well as other invertebrates). These highlight the role that **reciprocal inhibition** plays in regulating the phasic activity of locomotory muscles. First, because muscles can only shorten when developing force, the reciprocating movements of any joint require that muscles be arranged in opposing sets of antagonists (e.g. flexors and extensors). This means that sensory feedback to antagonist muscles which act at a given joint and, more generally, across the joints of a limb acts in opposing fashion to stimulate one set of muscles while inhibiting the opposing set. For example, when a terrestrial vertebrate's hindlimb lands on the ground and its ankle initially begins to flex as the joint moment increases, the ankle extensors will be stretched as they begin to develop force, causing increased spindle Ia afferent feedback drive to the α-motorneuron pool of the ankle extensors (Fig. 10.4). As discussed above, this occurs via a monosynaptic pathway within spinal cord segments (lumbar vertebrae 2, 3 and 4 in humans and some other mammals) which contribute sensory and motor innervation to the muscle. In addition to their excitatory feedback to the muscle's own motor neurons, the Ia afferents of the ankle extensors also synapse onto interneurons located within the spinal cord which, in turn, inhibit the α-motoneurons of the ankle *flexor* muscles. This ensures that activation of the ankle extensors resulting from stretch due to ankle flexion prevents the ankle flexors from being activated. At the end of support, when the ankle must be flexed and the hindlimb swung forward, the phase and pattern of motor stimulation is reversed. Activation of the ankle (and other limb) flexors results in inhibitory feedback of the ankle extensors via interneurons within the spinal cord. Thus there is reciprocal inhibition via spinal cord interneurons between the extensors and flexors within the same limb. This same general pattern of reciprocal inhibition of flexors versus extensors within a limb also holds for invertebrates and other forms of locomotion.

The second feature of locomotion is that contralateral limbs often operate in reciprocal fashion (Fig. 10.8) (see also Chapter 3). This is true for walking, running and trotting gaits (although it is not true for the hopping and bounding gaits of vertebrates). It is also true for undulatory swimming (in which opposite sides of the body axis are activated out of phase) but is not true for pectoral fin swimming or for flying. We shall concentrate here on the out-of-phase motions of the limbs associated with gaits commonly used by terrestrial animals. The reciprocal motion of the limbs means that when one limb is in contact with the ground and supporting body weight, its extensors must be activated. At this time, the opposing limb is being swung forward by its flexors to anticipate the next support phase. The reciprocal out-of-phase timing of the stance phase extensor muscles and the swing phase flexor muscles is mediated by reciprocal inhibition within the appropriate spinal cord segments (Fig. 10.4). Again, using the ankle extensors and flexors of terrestrial vertebrates as an example, activation of the ankle extensors in the stance phase limb results not only in spindle feedback for enhanced motor drive to the same agonist muscles and inhibitory feedback to the ankle flexor antagonists, but also feedback to muscles of the contralateral swing phase limb via interneurons which cross between the left and right motor neuron pools of the body. Once again, this same pattern holds for invertebrates (Figs. 10.6 and 10.7). In our example of terrestrial vertebrate ankle extensors, this contralateral feedback exerts mainly an *inhibitory* influence on the *extensors* of the opposing limb, but may also provide an *excitatory* influence on the *flexors* of the opposing swing phase limb (Fig. 10.8). Thus there is a second level of reciprocal inhibition which occurs via interneurons that transmit sensory feedback to muscles of contralateral limbs.

10.5 Distributed control: the role of central pattern generators

The coordinated timing of muscle extensors and flexors within and between limbs allows for more distributed control of muscle function which simplifies the command requirements of the central nervous system of both insects and vertebrates. Much of an animal's regulation of its limb movement patterns as it maneuvers or changes speed and gait is largely accomplished as an unconscious act, reflecting local control for meeting the mechanical requirements for body support and movement. This allows the central nervous system to be attentive to other functions and needs of the animal. Although the motor response of an animal to the sight, sound or smell of a predatory threat involves the integration of sensory stimuli via the central nervous system to organize and plan the animal's locomotive response, once initiated, it is largely controlled at local levels via the reflex feedback pathways described above.

Control of the relative timing and strength of flexor–extensor activity within and between limbs is believed to be mediated by networks of neurons called **central pattern generators** (CPGs). CPGs represent clusters of nerve cells located within the spinal cord of vertebrates or nervous system ganglia of insects which have *rhythmic* burst-generating properties. The organization and network properties of such CPGs are rarely identified in any discrete fashion, but models for their organization can be constructed which accurately describe their motor output and their response to changes in sensory input. The evidence for the existence for CPGs rests largely on experimental observations of animals in which coordinated movement patterns of the limbs (cats, turtles, cockroaches, and locusts) or undulation of the body axis (lampreys and dogfish) can be initiated and maintained independent of any functional link to higher brain centers.

One such model of a CPG is depicted in Fig. 10.9(a), which shows a 'flexor-burst generator' model developed by Pearson (1976) to describe how the reciprocating activity of flexors and extensors within the limbs of walking cockroaches is achieved. In this model, four interneurons within the nervous system are hypothesized to interact in such away that they produce a oscillating change in the membrane potential of a key interneuron (interneuron 1). The output of this network via interneuron 1 exerts a reciprocal excitatory input to the flexor motor neurons and inhibitory input (via another interneuron) to the extensor motor neurons of the limb. Descending central motor commands from the brain and 'higher centers' (to initiate and sustain movement) are believed to exert an excitatory effect on the flexor-burst CPG (establishing the frequency and amplitude of its oscillatory output) which drives the extensor motor neurons. Hence, in this model, inhibition of the extensor motor neurons (and muscles) is achieved via the inhibitory input from the CPG at the time when the flexor motor neurons are being activated, evident by their firing a series of action potentials to activate the flexor muscles.

A similar 'half-center model' has also been developed to describe the rhythmic reciprocal pattern of flexor and extensor motor neurons in the hindlimb of a cat (Fig. 10.9(b)). In contrast with the asymmetric flexor-burst generator model used to describe the control of the cockroach limb, the half-center model is symmetric, with mutually excitatory and inhibitory components to control the out-of-phase activity of flexors and extensors. In the half-center model the putative interneurons which drive the CPG located within the spinal cord are linked with their associated flexor or extensor motor neurons to form a 'half-center'. Reciprocal inhibition via interneurons within the CPG network and between flexor and extensor motor neurons results in an out-of-phase activation of the flexor and extensor muscles of the limb. Like the CPG network of the cockroach, the cat CPG model relies on descending motor commands from higher (brain) centers to excite the half-center network and establish a basic frequency and magnitude of oscillatory change in membrane potential to drive the alternating activation of the flexors and extensors within the limb. The mutual inhibition of each half-center facilitates the maintained rhythm of the motor pattern once it is established. Other more complicated CPG models have been developed. However, except for the most simple invertebrate motor systems, the exact pathways and neurons involved

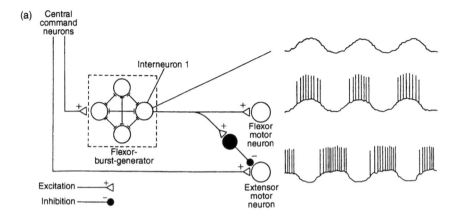

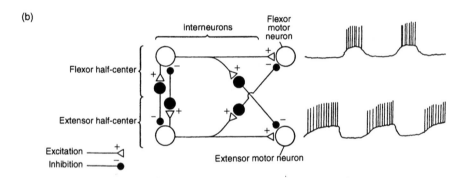

Fig. 10.9 Two hypothetical neuronal circuits which comprise a CPG network. (a) A 'flexor-burst-generator' model to describe how the reciprocating activity of flexors and extensors within the limbs of walking cockroaches is achieved (via oscillations in the membrane potential of Interneuron 1). This CPG network lies within one of the thoracic motor ganglia of the cockroach. (b) A 'half-center model' developed to describe the rhythmic reciprocal pattern of flexor and extensor motor neurons in the hindlimb of a cat. The cat CPG is believed to reside within the spinal cord. Both CPG networks rely on recriprocal inhibition mediated via interneurons to establish out-of-phase activation of motor neurons that excite flexor and extensor muscles within the limb. Central commands from the brain can modulate CPG output and hence locomotor behavior in response to sensory input, such as visual, auditory or olfactory cues (from Pearson 1976, with permission from Scientific American, Inc.).

in such pattern-generating networks are still not well known. A key principle of these networks is that they do not rely on a pacemaker cell to establish and maintain a motor rhythm as occurs in a beating heart. Instead, external sensory input, often integrated with reflex feedback resulting from the motor activation of the muscles themselves, facilitates the rhythmicity and pattern established by the spinal CPG, in association with descending commands via the spinal cord from higher central nervous system levels.

The alternating rhythmicity of antagonist muscles, such as flexors and extensors, within a limb associated with swing and support phases of a locomotor cycle can readily be explained by the simple CPG network models as those described above. However, the existence of a CPG for each limb does not mean that sensory information is unimportant. Indeed, the ability to adjust limb-movement patterns to accommodate irregularities in the environment is an ongoing requirement of stable coordinated locomotion. In fact, the timing of muscle activation patterns which result in shifts in the relative onset, offset and duration of motor activation within a limb must be modulated by sensory feedback to the CPG, as well as directly to the motor neuron pools which supply the muscles of a limb. Exactly how such shifts in timing are mediated among functional agonists and antagonists is less well known.

Because there is little change in the duration of the swing phase when animals move at different speeds, decreases in the duration of the support phase underlie faster speed movement. This suggests that completion of the stance phase triggers activation of the next swing phase (Pearson 1976). In cats, two conditions appear to be necessary for this: the hip must be extended, and the extensor muscles must be unloaded. The principle of limb unloading to initiate the swing phase of the stride also applies to the stepping of cockroaches. This is borne out by the observation that the campaniform sensilla, which detect the strains in the cuticle resulting from extensor activity during stance, inhibit the flexor-burst generating system of interneurons (Pearson 1976). As the leg is extended and becomes unloaded toward the end of stance, this inhibition is lost (cuticle strains decrease), facilitating activation of the limb flexors and the initiation of the limb's swing phase. The switching from swing to stance is also subsequently initiated by sensory input. In this case, the hair receptors on the leg, which are activated by the rapid motion of the leg, inhibit the flexor CPG and excite the extensor motor neurons. This causes the flexors to relax and the extensors to contract, initiating stance. Moreover, sensory feedback from the hair cells ensures that the position of the limb at the start of stance is always very nearly the same, independent of the distance that it traveled backward during the previous step.

Another example of how sensory feedback modulates the motor output to the limb is displayed when the top surface of an animal's foot is stimulated. Activation of proprioceptors in the skin overlying the foot causes the animal (observed in a cat) to flex its limb more strongly, lifting the foot higher during the swing phase (Pearson 1976). This reflexive elevation of the limb and foot enables an animal to step over obstacles that it may encounter in its environment. Therefore reflexive sensory feedback has at least two important functions in controlling phasic motor activity during locomotion. The first is to switch the motor program from one phase to the other (e.g. from swing to stance), and the second is to modulate the motor output within a single phase of the limb's movement.

The relative timing of limb-movement patterns relative to one another must also engage CPGs controlling pairs of limbs, for example when an animal changes speed or gait. Hence, higher-level networks within the nervous system must exist to facilitate the coordinated timing of movement patterns among multiple limbs of the animal, maintaining their appropriate phase of activation. The reciprocal inhibition provided by interneurons to opposing contralateral motor unit pools of a pair of limbs (e.g. the hindlimbs of a dog or a cockroach) necessarily represents a key component of local CPG control for the out-of-phase relationship of the paired limb movement. However, for those gaits (such as hopping or bounding) or modes of locomotion (such as flight) which require synchronous activation of paired appendages, it is clear that the intrinsic pattern of reciprocal inhibition must be overridden (when an animal changes gait) or have been lost through the evolution of a new network which ensures paired synchrony of limb movement patterns. Owing to the greater complexity of central nervous system organization and function, the anatomical and neurophysiological basis for such shifts in limb movement patterns remains poorly understood.

10.6   Two case studies of motor control

Two examples of animals which use different modes of locomotion provide interesting and compelling case studies of neuromotor control and how changes in muscle recruitment pattern accommodate changes in locomotive performance. The first examines how a fly's underlying neuromuscular properties and reflex pathways are organized to mediate and control its maneuvering ability during flight. The second shows how neuromuscular patterns of recruitment may be adjusted to compensate for physiological effects of temperature on the swimming performance of fish.

10.6.1   Visual motor control of fly flight

In addition to their wings, dipterous flies possess a pair of equilibrium organs termed 'halteres' which have evolved from the hindwings of their ancestors. As a result, the halteres are equipped with a similar organization of steering and flight muscles to those that control motions of the fly's (fore)wings. The haltere muscles are also under direct control of motion-sensitive visual motorneurons (Fig. 10.10). This provides a means for visual motor control that is linked via the halteres to the wings, which are ultimately responsible for generating the aerodynamic forces necessary for controlling the fly's flight path. In response to visual stimuli, or experimental perturbations of its visual world (this can be

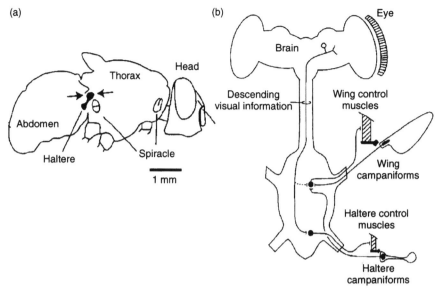

Fig. 10.10 (a) The haltere, a modified ancestral hindwing, of flies which functions like a gyroscope to stabilize the fly's body by beating back-and-forth antiphase to the beating of the wings. (b) The neural control circuit by which inertial and motor movements of the halteres is sensed and fed forward to control the motor output to the wing's steering, muscles. Visual sensory input from the fly's eyes is integrated with the efferent control from the halteres, as well as the afferent sensory input from the campaniform sensilla of the wings. See text for additional detail. (Reproduced from Chan *et al.* (1998), with permission from the American Association for the Advancement of Science.)

done by controlling the relative equatorial motion of vertical bands of light that the fly sees while it is tethered to a force transducer), a fly will alter (or attempt to alter, when tethered) its body orientation to change its flight direction. Hence, visual stimuli strongly influence a fly's flight behavior, and the halteres are key to this sensorimotor integration.

During flight, the club-shaped halteres beat out of phase with respect to the beating forewings. Although the halteres have lost their aerodynamic function, their sensory components are enhanced relative to those of the forewings. For example, in a blowfly (*Calliphora*) each haltere is equipped with 335 campaniform sensilla organized in distinct fields at the base of the haltere. Sensory neurons innervating these campaniforms (Fig. 10.3) encode Coriolis forces which result from the cross-product of the haltere's linear velocity with the angular velocity of the fly's body about its pitch, yaw and roll axes. By projecting directly onto the motor neurons which drive the forewings during flight, the campaniform sensory nerves provide stabilizing control for flight (Chan *et al.* 1998). This connection represents a monosynaptic pathway similar to vertebrate spindle Ia afferents, providing a rapid response of wing movements to perturbations of the motions of the halteres. If their halteres are removed, flies are unstable and soon crash to the ground. The halteres represent an *efferent* control system, in the sense that the motor output from the haltere muscles influences the sensory output of the campaniforms at their base. This information is then fed forward to the motor neurons of the forewings. Once again, this is similar to the adjustment of muscle spindle sensitivity to length change by means of the fusimotor efferent nerves (γ) regulating the length of intrafusal fibers (section 10.2).

The maneuvering ability of flies is quite exceptional and is well displayed when a male fly tracks a female fly during mating flight behavior (Wagner 1986).This ability depends on the stabilizing control that the halteres provide via their integration of visual stimuli with mechanosensory control of the wings. However, this raises the problem of how this compensatory control can be overridden when a fly seeks to make a voluntary maneuver. One possibility favored by Chan *et al.* (1998) is that the fly may actively manipulate the properties or sensitivity of its haltere-based reflex loop (Fig. 10.10) to generate voluntary maneuvers. This might be accomplished by activating the steering control muscles of the haltere. This would alter the haltere's kinematics, modulating the afferent output of its campaniform sense organs to the motor neurons controlling the wing steering muscles. As a result, 'compensatory' motions of the wings would be activated, causing them to induce a change in flight direction via pitch, roll and/or yaw of the fly's body. In effect, this suggests that voluntary control is accomplished by 'tricking' the fly's flight steering muscles into sensing that they must compensate for a perturbation which does not really exist. Thus the ability of the wing steering muscles to respond effectively and rapidly to actual perturbations sensed by the halteres would provide equally effective control of voluntary maneuvers.

10.6.2 Fish swimming: motor recruitment in relation to temperature

As ectotherms, most fish encounter seasonal and even daily changes in water temperature. This requires that their axial musculature also operate over a range of temperature. Because of the thermal effects on reaction rate, muscles contract more slowly at lower temperatures. In general, a two- to three-fold decline in contraction speed can be expected for a 10 °C decrease in temperature. An interesting consequence of this is that fish which experience shifts in environmental temperature alter the recruitment pattern of their slow red (SO) and fast white (FG) axial muscle fibers in response to temperature changes in order to maintain a more uniform swimming capacity (Rome *et al.* 1984). At 20 °C, swimming carp are able to sustain a cruising speed of 0.45 m/s powered exclusively by their red muscle fibers (Fig. 10.11(a)). When the water

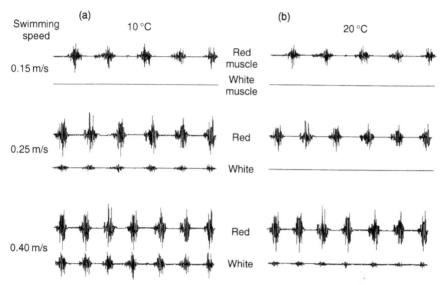

Fig. 10.11 EMGs recorded from the red and white axial musculature (see Fig. 4.14) of a carp swimming at three different speeds and at two different temperatures. To swim faster the carp (and other fish) must recruit more muscle and progressively shift its recruitment from red muscle fibers to white muscle fibers. At a higher temperature (20 °C), it can sustain swimming speeds up to 0.35 m/s by recruiting only its red muscle. At 0.40 m/s it must begin to recruit its faster white muscle fibers to sustain faster swimming speeds. When the fish encounters colder water (10 °C), it begins to recruit its white muscle fibers, in addition to its red fibers, at a slower speed (0.25 m/s). This is because its muscle fibers contract more slowly and produce work at a slower rate at the colder temperature. This 'compression' of the fish's muscle fiber recruitment order represents a motor strategy by which ectothermic animals, such as carp, can sustain comparable levels of performance in the face of acute changes in temperature regime. (Adapted from Rome et al. (1984).)

temperature is lowered to 10 °C, carp begin to recruit their white musculature at 0.25 m/s, in addition to their red muscle fibers (Fig. 10.11(b)). The same pattern of muscle recruitment is observed at the two temperatures, but it is 'compressed' into a narrower speed range at the lower temperature.

This shift in motor recruitment in response to a change in water (and body) temperature enables the fish to compensate for the slower rates of myosin-mediated cross-bridge cycling due to a decrease in temperature. By recruiting faster-contracting white fibers at lower temperatures, the carp are able to achieve a similar contraction speed as their red fibers at the higher temperature, allowing them to sustain a similar swimming speed. Although this enables many fish to compensate for and maintain performance in response to changes in environmental temperature, it is likely that the increased reliance on white glycolytic fibers reduces their capacity for swimming endurance at the colder temperature. Certainly the top speeds of fish are limited by temperature. Does this pose a significant ecological threat to the fish? Possibly not, given that its prey and other ectothermic predators experience the same temperature-slowing effects on locomotive performance. Of course, predators capable of maintaining elevated muscle, brain and eye temperatures (billfish, sea mammals and other swimming mammals) probably have a distinct edge!

Similar patterns of temperature-dependent shifts in motor unit recruitment have been observed in other fish, as well as certain other lizards, but it is not a pattern that can necessarily be readily generalized to all ectothermic taxa. Nevertheless, it is a clever way by which some ectotherms are able to compensate for temperature-dependent effects on muscle contractile function and certainly it is an area of research that deserves additional study.

10.7 Summary

In this chapter, we have reviewed the basic sensory elements of vertebrates and insects which provide local feedback control to the muscles which operate their limbs (or body axis). These are important to coordinated and stable movement. In vertebrates, these involve muscle spindle organs which sense length changes of a muscle, Golgi tendon organs which sense the force transmitted to a muscle's tendon, and various types of proprioceptors which sense pressure at the body surface and within joints. In insects, these involve hair-like exteroreceptors which sense tactile stimuli at the surface of the animal and mechanosensitive proprioreceptors which exist either at the cuticle surface or within the joints of the animal. These sensory elements and their reflex pathways to the motor neurons which innervate the muscles provide local sensorimotor integration that is critical to decentralized control of locomotive movements.

In both vertebrates and insects, direct monosynaptic pathways exist which facilitate rapid motor responses to sensory stimuli, allowing animals to adjust rapidly to their environment when moving. This involves excitatory input to motor units which activate the muscles increasing their force output. In vertebrates this is classically represented as a 'stretch reflex' of the spindles within the muscle that enhances the motor drive to a muscle to resist its being stretched. However, whereas motor recruitment in vertebrates involves the activation of discrete motor units in an all-or-none fashion, in insects and other invertebrates recruitment is via graded junction potentials which sum along fibers that are multiply innervated by different motor nerves. In addition, whereas vertebrate muscles are typically innervated by many motor nerves and comprise many motor units, invertebrate muscles are typically innervated by only a few motor nerves. Motor recruitment in both groups is generally from slow to faster contracting fibers within the muscle. In vertebrates the progressive recruitment from small slow (oxidative) to intermediate to large fast (glycolytic) units is termed the 'size principle' and is largely considered an inviolate ordering of motor recruitment.

Sensorimotor integration and control of muscle activation within and between limbs also involves pathways mediated via interneurons within the spinal cord of vertebrates and motor ganglia of insects which control the out-of-phase activation of flexors and extensors. This reflects the fundamental role that reciprocal inhibition plays in mediating the relative phase of muscle activation associated with the swing and stance phases of a limb. The rhythmic timing of limb- and body-movement patterns is mediated by CPGs which constitute networks of neurons that reside at local levels within the spinal cord or motor ganglia. CPGs facilitate the decentralized control of basic motor patterns and locomotive movement. Higher centers which involve the brain and sensory reception to the central nervous system provide 'descending' input to local CPGs to initiate and control more complex motor behaviors. However, the ability of local CPGs to maintain rhythmic motor behaviors that can be modulated according to local sensory feedback within and between limbs is fundamental to coordinated stable movement of the animal and is also key to the changing leg-movement patterns associated with changes of gait and running speed. Future work which incorporates biomechanical analysis of the design and properties of animal limbs, together with electrophysiological studies and computational modeling of the properties of neural networks, will provide an improved understanding of the remarkable capacity of animals for moving over broad ranges of speed and maneuvering through complex environments.

References

Further reading for each chapter

Chapter 1

Alexander, R.M. (1983). *Animal mechanics* (2nd edn). Blackwell Scientific, London.
Wainwright, S.A., Biggs, W.D., Currey, J.D. and Gosline, J.M. (1976). *Mechanical design in organisms*. Arnold, London.

Chapter 2

Currey, J.D. (1984). *The mechanical adaptations of bone*. Princeton University Press.
Lieber, R.L. (1992). *Skeletal muscle structure and function*. Williams and Wilkins, Baltimore, MD.
Wainwright, S.A., Biggs, W.D., Currey, J.D. and Gosline, J.M. (1976). *Mechanical design in organisms*. Arnold, London.

Chapter 3

Biewener, A.A. (1990). Biomechanics of mammalian terrestrial locomotion. *Science*, **250**, 1097–1103.
Hildebrand, M.B. (1988). *Analysis of vertebrate structure* (3rd edn). Wiley, New York.
Muybridge, E. (1957). *Animals in motion* (ed. L.S. Brown). Dover Publications, New York.

Chapter 4

Videler, J.J. (1993). *Fish swimming*. Chapman & Hall, London.

Vogel, S. (1994). *Life in moving fluids. The physical biology of flow.* Princeton University Press.

Chapter 5

Dudley, R. (2000). *The biomechanics of insect flight. Form, function, evolution.* Princeton University Press.
Heinrich, B. (1993). *The hot-blooded insects: strategies and mechanisms of thermoregulation.* Harvard University Press, Cambridge, MA.
Norberg, U.M. (1990). *Vertebrate flight.* In *Zoophysiology*, Vol. 27. Springer-Verlag, New York.
Vogel, S. (1994). *Life in moving fluids. The physical biology of flow.* Princeton University Press.

Chapter 6

Mitchison, T.J. and Cramer, L.P. (1996). Actin-based cell motility and cell locomotion. *Cell*, **84**, 371–9.
Stossel, T.P. (1994). The machinery of cell crawling. *Scientific American*, **9**, 54–63.
Theriot, J.A. and Mitchison, T.J. (1991). Actin microfilament dynamics in locomoting cells. *Nature*, **352**, 126–31.

Chapter 7

Aerts, P. (1998). Vertical jumping in *Galago senegalensis*: the quest for an obligate power amplifier. *Philosophical Transactions of the Royal Society of London, Series B*, **353**, 1607–20.
Alexander, R.M. (1988). *Elastic mechanisms in animal movement.* Cambridge University Press.
Autumn, K., Liang, Y.A., Hsieh, S.T., *et al.* (2000). Adhesive force of a single gecko foot-hair. *Nature*, **405**, 681–5.
Cartmill, M. (1985). Climbing. In *Functional vertebrate morphology* (ed. M. Hildebrand, D.M. Bramble, K.F. Liem and D.B. Wake), pp. 73–88. Harvard University Press, Cambridge, MA.
Marsh, R.L. (1994). Jumping ability of anuran amphibians. *Advances in Veterinary Science and Comparative Medicine*, **38B**, 51–111.

Chapter 8

Alexander, R.M. (1999). *Energy for animal life.* Oxford University Press.
Brooks, G.A., Fahey, T.D. and White, T.G. (1996). *Exercise physiology: human bioenergetics and its applications.*

Gleeson, T.T. (1991). Patterns of metabolic recovery from exercise in amphibians and reptiles. *Journal of Experimental Biology*, **160**, 187–207.

Hochachka, P.W. (1994). Solving the common problem: matching ATP synthesis to ATP demand during exercise. *Advances in Veterinary Science and Comparative Medicine*, **38A**, 41–54.

Taylor, C.R. (1994). Relating mechanics and energetics during exercise. *Advances in Veterinary Science and Comparative Medicine*, **38A**, 181–215.

Weibel, E.R. (1984). *The Pathway for Oxygen*. Harvard University Press, Cambridge, MA.

Chapter 9

Alexander, R.M. (1999). *Energy for animal life*. Oxford University Press.

Bennett, A.F. (1994). Exercise performance of reptiles. *Advances in Veterinary Science and Comparative Medicine*, **38B**, 113–38.

Taylor, C.R. (1994). Relating mechanics and energetics during exercise. *Advances in Veterinary Science and Comparative Medicine*, **38A**, 181–215.

Chapter 10

Goslow, G.E., Jr (1985). Neural control of locomotion. In *Functional vertebrate morphology* (ed. M. Hildebrand, D.M. Bramble, K.F. Liem and D.B. Wake). Harvard University Press, Cambridge, MA.

Grillner, S. (1985). Neurobiological bases of rhythmic motor acts in vertebrates. *Science*, **228**, 143–9.

Pearson, K.G. (1976). The control of walking. *Scientific American*, **235**, 72–86.

Pearson, K.G. (2000). Neural adaptation in the generation of rhythmic behavior. *Annual Review of Physiology*, **62**, 723–53.

References

Aerts, P. (1998). Vertical jumping in *Galago senegalensis*: the quest for an obligate power amplifier. *Philosophical Transactions of the Royal Society of London, Series B*, **353**, 1607–20.

Alexander, R.M. (1976). Estimates of speeds of dinosaurs. *Nature*, **261**, 129–30.

Alexander, R.M. (1981). Factors of safety in the structure of animals. *Science Progress*, **67**, 119–40.

Alexander, R.M. (1983). *Animal mechanics* (2nd edn). Blackwell Scientific, London.

Alexander, R.M. (1984). Stride length and speed for adults, children, and fossil hominids. *American Journal of Physical Anthropology*, **63**, 23–7.

Alexander, R.M. (1988). *Elastic mechanisms in animal movement*. Cambridge University Press.

Alexander, R.M. and Jayes, A.S. (1983). A dynamic similarity hypothesis for the gaits of quadrupedal mammals. *Journal of Zoology*, 201, 135–52.

Altenbach, J.S. and Hermanson, J.W. (1987). Bat flight muscle function and the scapulohumeral lock. In *Recent advances in the study of bats* (ed. M.B. Fenton, P.A. Racey and J.M.V. Rayner), pp. 100–18. Cambridge University Press.

Autumn, K., Liang, Y.A., Hsieh, S.T., *et al.* (2000). Adhesive force of a single gecko foot-hair. *Nature*, 405, 681–5.

Baker, E.J. and Gleeson, T.T. (2001). The effects of intensity on the energetics of brief locomotor activity. *Journal of Experimental Biology*, 202, 3081–7.

Baudinette, R.V. and Biewener, A.A. (1998). Young wallabies get a free ride. *Nature*, 395, 653–4.

Baudinette, R.V., Snyder, G.K. and Frappell, P.B. (1992). Energetic cost of locomotion in the tammar wallaby. *American Journal of Physiology*, 262, 771–78.

Beer, F.P. and Johnston, E.R. (1981). *Mechanics of materials*. McGraw-Hill, New York.

Bennett, A.F. (1978). Activity metabolism of the lower vertebrates. *Annual Review of Physiology*, 40, 447–69.

Bennet, M.B., Ker, R.F., Dimery, N.J. and Alexander, R.M. (1986). Mechanical properties of various mammalian tendons. *Journal of the Zoological Society of London*, 209, 537–48.

Bennet-Clark, H.C. (1975). The energetics of the jump of the locust, *Schistocerca gregaria*. *Journal of Experimental Biology*, 63, 53–83.

Bennet-Clark, H.C. (1977). Scale effects in jumping animals. In *Scale effects in animal locomotion* (ed. T. J. Pedley), pp. 185–201. Academic Press, London.

Bennet-Clark, H.C. and Lucey, E.C.A. (1967). The jump of the flea: a study of the energetics and a model of the mechanism. *Journal of Experimental Biology*, 47, 59–76.

Berg, H.C. (1983). *Random walks in biology*. Princeton University Press.

Bertram, J.E.A., Ruina, A., Cannon, C.E., Chang, Y.H. and Coleman, M.J. (1999). A point-mass model of gibbon locomotion. *Journal of Experimental Biology*, 202, 2609–17.

Biewener, A.A. (1989). Scaling body support in mammals: limb posture and muscle mechanics. *Science*, 245, 45–8.

Biewener, A.A. (1990). Biomechanics of mammalian terrestrial locomotion. *Science*, 250, 1097–1103.

Biewener, A.A. (1998). Muscle–tendon stresses and elastic energy storage during locomotion in the horse. *Comparative Biochemistry and Physiology B*, 120, 73–87.

Biewener, A.A. and Baudinette, R.V. (1995). *In vivo* muscle force and elastic energy storage during steady-speed hopping of tammar wallabies (*Macropus eugenii*). *Journal of Experimental Biology*, 198, 1829–41.

Biewener, A.A. and Blickhan, R. (1988). Kangaroo rat locomotion: design for elastic energy storage or acceleration? *Journal of Experimental Biology*, **140**, 243–55.

Biewener, A.A. and Dial, K.P. (1995). *In vivo* strain in the humerus of pigeons (*Columba livia*) during flight. *Journal of Morphology*, **225**, 61–75.

Biewener, A.A., Alexander, R.M. and Heglund, N.C. (1981). Elastic energy storage in the hopping of kangaroo rats (*Dipodomys spectabilis*). *Journal of Zoology*, **195**, 369–83.

Biewener, A.A., Dial, K.P. and Goslow, G.E., Jr (1992). Pectoralis muscle force and power output during flight in the starling. *Journal of Experimental Biology*, **164**, 1–18.

Biewener, A.A., Konieczynski, D.D. and Baudinette, R.V. (1998a). *In vivo* muscle force–length behavior during steady-speed hopping in tammar wallabies. *Journal of Experimental Biology*, **201**, 1681–94.

Biewener, A.A., Corning, W.R. and Tobalske, B.T. (1998b). *In vivo* pectoralis muscle force—length behavior during level flight in pigeons (*Columba livia*). *Journal of Experimental Biology*, **201**, 3293–307.

Blake, R.W. (1981). Mechanics of drag-based mechanisms of propulsion in aquatic vertebrates. *Symposia of the Zoological Society of London*, **48**, 29–52.

Blickhan, R. and Full, R.J. (1987). Locomotion energetics of the ghost crab. II: Mechanics of the centre of mass during walking and running. *Journal of Experimental Biology*, **130**, 155–74.

Block, B.A. (1986) Structure of the brain and eye heater tissue in marlins, sailfish, and spearfishes. *Journal of Morphology*, **190**, 169–89.

Block, B.A. (1994). Thermogenesis in muscle. *Annual Review of Physiology*, **56**, 535–77.

Brett, J.R. (1965). The relation of size to rate of oxygen consumption and sustained swimming speed of sockeye salmon (*Oncorhynchus nerka*). *Journal of the Fisheries Research Board of Canada*, **22**, 1491–1501.

Bundle, M.W., Hoppeler, H., Vock, R., Tester, J.M. and Weyand, P.G. (1999). High metabolic rates in running birds. *Nature*, **397**, 31–32.

Burrows, M. (1989). Processing of mechanosensory signals in local reflex pathways of the locust. *Journal of Experimental Biology*, **146**, 209–228.

Burrows, M. (1996). Controlling local movements of the legs. In *The neurobiology of the insect brain* (ed. M. Burrows), pp. 254–325. Oxford University Press.

Burrows, M. and Hoyle, G. (1973). The mechanism of rapid running in the ghost crab, *Ocypode ceratophthalma*. *Journal of Experimental Biology*, **58**, 327–49.

Burrows, M. and Pfluger, H.J. (1995). Action of locust neuromodulatory neurons is coupled to specific motor patterns. *Journal of Neurophysiology*, **74**, 347–57.

Carey, F.G. (1973). Fishes with warm bodies. *Scientific American*, **228**, 36–44.

Cartmill, M. (1979). The volar skin of primates: its frictional characteristics and their functional significance. *American Journal of Physical Anthropology*, **50**, 497–510.

Cartmill, M. (1985). Climbing. In *Functional vertebrate morphology* (ed. M. Hildebrand, D.M. Bramble, K.F. Liem and D.B. Wake), pp. 73–88. Harvard University Press, Cambridge, MA.

Cavagna, G.A. and Kaneko, M. (1977). Mechanical work and efficiency in level walking and running. *Journal of Physiology*, **268**, 467–81.

Cavagna, G.A., Heglund, N.C. and Taylor, C.R. (1977). Mechanical work in terrestrial locomotion: two basic mechanisms for minimizing energy expenditures. *American Journal of Physiology*, **233**, R243–61.

Chan, W.P., Prete, F. and Dickinson, M.H. (1998). Visual input to the efferent control system of a fly's 'gyroscope'. *Science*, **280**, 289–92.

Chang, Y.H., Bertram, J.E.A. and Ruina, A. (1997). A dynamic force and moment analysis system for brachiation. *Journal of Experimental Biology*, **200**, 3013–20.

Currey, J.D. (1981). What is bone for? Property–function relationships in bone. In *Mechanical properties of bone* (ed. S. C. Cowin), pp. 13–26. ASME, New York.

Currey, J.D. (1984). *The mechanical adaptations of bone*. Princeton University Press.

Daniel, T., Jordan, C. and Brunbaum, D. (1992). Hydromechanics of swimming. In *Advances in comparative environmental physiology. 11: Mechanics of locomotion* (ed. R.M. Alexander), pp. 17–49. Springer-Verlag, Berlin.

Daniel, T.L. (1984). Unsteady aspects of aquatic locomotion. *American Zoologist*, **24**, 121–34.

Dawson, T.J. and Taylor, C.R. (1973). Energetic cost of locomotion in kangaroos. *Nature*, **246**, 313–14.

DeMont, M.E. and Gosline, J.M. (1988). Mechanics of jet propulsion in the hydromedusan jellyfish, *Polyorchis penicillatus*. I: Mechanical properties of the locomotor structure. *Journal of Experimental Biology*, **134**, 313–32.

Dewar, H. and Graham, J.B. (1994). Studies of tropical tuna swimming performance in a large water tunnel—energetics. *Journal of Experimental Biology*, **192**, 13–31.

Dial, K.P. (1992a). Activity patterns of the wing muscles of the pigeon (*Columba livia*) during different modes of flight. *Journal of Experimental Zoology*, **262**, 357–73.

Dial, K.P. (1992b). Avian forelimb muscles and nonsteady flight. Can birds fly without using the muscles of their wings? *AukD*, **109**, 874–85.

Dial, K.P. (2003). Wing-assisted incline running and the evolution of flight. *Science*, **299**, 402–404.

Dial, K.P. and Biewener, A.A. (1993). Pectoralis muscle force and power output during different modes of flight in pigeons (*Columba livia*). *Journal of Experimental Biology*, **176**, 31–54.

Dial, K.P., Biewener, A.A., Tobalske, B.W. and Warrick, D.R. (1997). Direct assessment of mechanical power output of a bird in flight. *Nature*, **390**, 67–70.

Dickinson, M.H. (1996). Unsteady mechanisms of force generation in aquatic and aerial locomotion. *American Zoologist*, **36**, 537–54.

Dickinson, M.H., Lehmann, F.-O. and Sane, S.P. (1999). Wing rotation and the aerodynamic basis of insect flight. *Science*, **284**, 1954–60.

Douglas, M.M. (1981). Thermoregulatory significance of thoracic lobes in the evolution of insect wings. *Science*, **211**, 84–6.

Drucker, E.G. and Lauder, G.V. (2000). A hydrodynamic analysis of fish swimming speed: wake structure and locomotor force in slow and fast labriform swimmers. *Journal of Experimental Biology*, **203**, 2379–93.

Dudley, R. (2000). *The biomechanics of insect flight. Form, function, evolution.* Princeton University Press.

Edwards, E.B. and Gleeson, T.T. (2001). Can energetic expenditure be minimized by performing activity intermittently? *Journal of Experimental Biology*, **204**, 585–97.

Ellington, C.P. (1984). The aerodynamics of hovering insect flight. IV: Aerodynamic mechanisms. *Philosophical Transactions of the Royal Society of London, Series B*, **305**, 79–113.

Ellington, C.P. (1991). Limitations on animal flight performance. *Journal of Experimental Biology*, **160**, 71–91.

Ellington, C.P., van den Berg, C., Willmott, A.P. and Thomas, A.L.R. (1996). Leading-edge vortices in insect flight. *Nature*, **384**, 626–30.

Emerson, S.B. (1985). Jumping and leaping. In *Functional vertebrate morphology* (ed. M. Hildebrand, D.M. Bramble, K.F. Liem and D.B. Wake), pp. 58–72. Harvard University Press, Cambridge, MA.

Farley, C.T. and Taylor, C.R. (1991). A mechanical trigger for the trot–gallop transition in horses. *Science*, **253**, 306–8.

Farley, C.T., Glasheen, J. and McMahon, T.A. (1993). Running springs: speed and animal size. *Journal of Experimental Biology*, **185**, 71–86.

Ferry, L.A. and Lauder, G.V. (1996). Heterocercal tail function in leopard sharks: a three-dimensional kinematic analysis of two models. *Journal of Experimental Biology*, **199**, 2253–68.

Fish, F.E. (1996). Transitions from drag-based to lift-based propulsion in mammalian swimming. *American Zoologist*, **36**, 628–41.

Full, R.J. (1987). Locomotion energetics of the ghost crab. I. Metabolic cost and endurance. *Journal of Experimental Biology*, **130**, 137–53.

Full, R.J. (1991). Animal motility and gravity. *Physiologist*, **34**, S15–18.

Full, R.J. and Koditschek, D.E. (1999). Templates and anchors: neuromechanical hypotheses of legged locomotion on land. *Journal of Experimental Biology*, **202**, 3325–32.

Full, R.J. and Tu, M.S. (1991). The mechanics of a rapid running insect: two-, four- and six-legged locomotion. *Journal of Experimental Biology*, **156**, 215–31.

Full, R.J. and Tullis, A. (1990). Energetics of ascent: insects on inclines. *Journal of Experimental Biology*, **149**, 307–17.

Full, R.J., Anderson, B.D., Finnerty, C.M. and Feder, M.E. (1988). Exercising with and without lungs. I: The effects of metabolic cost, maximal oxygen transport and body size on terrestrial locomotion in salamander species. *Journal of Experimental Biology*, **138**, 471–85.

Full, R.J., Blickhan, R. and Ting, L.H. (1991). Leg design in hexapedal runners. *Journal of Experimental Biology*, **158**, 369–90.

Gaesser, G.A. and Brooks, G.A. (1984). Metabolic bases of excess post-exercise oxygen consumption: a review. *Medicine and Science in Sports*, **16**, 29–43.

Gambaryan, P. (1974). *How mammals run: anatomical adaptations*. Wiley, New York.

Garland, T.J. (1983). The relation between maximal running speed and body mass in terrestrial mammals. *Journal of Zoology*, **199**, 157–70.

Gehr, P., Mwangi, D.K., Ammann, A., Maloiy, G.M.O., Taylor, C.R. and Weibel, E.R. (1981). Design of the mammalian respiratory system. V: Scaling morphometric pulmonary diffusing capacity to body mass: wild and domestic mammals. *Respiration Physiology*, **44**, 61–86.

George, J.C. and Berger, A.J. (1966). *Avian myology*. Academic Press, New York.

Gibb, A.C., Dickson, K.A. and Lauder, G.V. (1999). Tail kinematics of the chub mackerel *Scomber japonicus*: testing the homocercal tail model of fish propulsion. *Journal of Experimental Biology*, **202**, 2433–47.

Gillis, G.B. (1996). Undulatory locomotion in elongate aquatic vertebrates: anquill: form locomotion since Sir James Gray. *American Zoolgist*, **36**, 656–65.

Glasheen, J.W. and McMahon, T.A. (1996). A hydrodynamic model of locomotion in the basilisk lizard. *Nature*, **380**, 340–2.

Gordon, A.M., Huxley, A.F. and Julian, F.J. (1966). The variation in isometric tension with sarcomere length in vertebrate muscle fibers. *Journal of Physiology*, **184**, 170–92.

Gordon, J.E. (1978). *Structures: or why things don't fall down*. Da Capo Press, New York.

Gunther, M.M. (1985). Biomechanische Voraussetzungen beim Absprung des Senegalgalagos. *Zeitschrift für Morphologie und Anthropologie*, **75**, 287–306.

Hall-Craggs, E.B.C. (1965). An analysis of the jump of the lesser galago (*Galago senegalensis*). *Journal of Zoology*, **147**, 20–9.

Hedrick, T.L., Tobalske, B.W. and Biewener, A.A. (2002). Estimates of circulation and gait change based on a three-dimensional kinetic analysis of flight in cockatiels (*Nymphicus hollandicus*) and ringed turtle-doves (*Streptopelia risoria*). *Journal of Experimental Biology*, **205**, 1389–1409.

Heglund, N.C. and Taylor, C.R. (1988). Speed, stride frequency and energy cost per stride. How do they change with body size and gait? *Journal of Experimental Biology*, **138**, 301–18.

Heglund, N.C., Fedak, M.A., Taylor, C.R. and Cavagna, G.A. (1982). Energetics and mechanics of terrestrial locomotion. IV: Total mechanical energy changes as a function of speed and body size in birds and mammals. *Journal of Experimental Biology*, **97**, 57–66.

Heinrich, B. (1993). *The hot-blooded insects: strategies and mechanisms of thermoregulation*. Harvard University Press, Cambridge, MA.

Heitler, W.J. and Burrows, M. (1977). The locust jump. I. The motor programme. *Journal of Experimental Biology*, **66**, 203–220.

Henneman, E. (1957). Relations between size of neurons and their susceptibility to discharge. *Science*, **126**, 1345–7.

Henneman, E., Somjen, G. and Carpenter, D.O. (1965). Excitability and inhibitability of motorneurons of different sizes. *Journal of Neurophysiology*, **28**, 599–620.

Hildebrand, M.B. (1988). *Analysis of vertebrate structure* (3rd edn). Wiley, New York.

Hill, A.V. (1950). The dimensions of animals and their muscular dynamics. *Science Progress*, **38**, 209–30.

Hoppeler, H., Mathieu, O. *et al.* (1981). Design of the mammalian respiratory system. VIII: Capillaries in skeletal muscles. *Respiration Physiology*, **44**, 129–50.

Houk, J.C. (1979). Regulation of stiffness by skeletomotor reflexes. *Annual Review of Physiology*, **41**, 99–114.

Hoyle, G. (1983). *Muscles and their neural control*. Wiley, New York.

Hutchinson, J.R. and Garcia, M. (2002). *Tyrannosaurus* was not a fast runner. *Nature*, **415**, 1018–21.

Jindrich, D.L. and Full, R.J. (1999). Many-legged maneuverability: dynamics of turning in hexapods. *Journal of Experimental Biology*, **202**, 1603–23.

Josephson, R.K., Malamud, J.G. and Stokes, D.R. (2000). Asynchronous muscle: a primer. *Journal of Experimental Biology*, **203**, 2713–22.

Ker, R.F., Bennett, M.B., Bibby, S.R., Kester, R.C. and Alexander, R.M. (1987). The spring in the arch of the human foot. *Nature*, **325**, 147–9.

Kier, W.M. (1996). Muscle development in squid: ultrastructural differentiation of a specialized muscle fiber type. *Journal of Morphology*, **229**, 271–88.

Kingsolver, J.G. and Koehl, M.A.R. (1994). Selective factors in the evolution of insect wings. *Annual Review of Entomology*, **39**, 415–51.

Knower, T., Shadwick, R.E., Katz, S.L., Graham, J.B. and Wardle, C.S. (1999). Red muscle activation patterns in yellowfin (*Thunnus albacares*) and skipjack (*Katsuwonus pelamis*) tunas during steady swimming. *Journal of Experimental Biology*, **202**, 2127–38.

Kooyman, G.L. and Ponganis, P.J. (1998). The physiological basis of diving to depth: birds and mammals. *Annual Review of Physiology*, **60**, 19–32.

Kram, R. (1997). Effect of reduced gravity on the preferred walk–run transition speed. *Journal of Experimental Biology*, **200**, 821–6.

LaBarbera, M. (1983). Why the wheels won't go. *American Naturalist*, **121**, 395–408.

Lindstedt, S.L., Hokanson, J.F., Wells, D.J., Swain, S.D., Hoppeler, H. and Navarro, V. (1991). Running energetics in pronghorn antelope. *Nature*, **353**, 748–750.

Loeb, G.E. and Gans, C. (1986). *Electromyography for experimentalists*. University of Chicago Press.

Lutz, G.J. and Rome, L.C. (1994). Built for jumping: the design of the frog muscular system. *Science*, 263, 370–2.

McMahon, T.A. and Cheng, G.C. (1990). The mechanics of running. How does stiffness couple with speed? *Journal of Biomechanics*, **23**, 65–78.

McMahon, T.A., Valiant, G. and Frederick, E.C. (1987). Groucho running. *Journal of Applied Physiology*, **62**, 2326–37.

Marden, J.H. and Kramer, M.G. (1994). Surface-skimming stoneflies: a possible intermediate stage in insect flight evolution. *Science*, **266**, 427–30.

Margaria, R. (1976). *Biomechanics and energetics of muscular exercise*. Oxford University Press.

Marsh, R.L. (1994). Jumping ability of anuran amphibians. *Advances in Veterinary Science and Comparative Medicine*, **38B**, 51–111.

Marsh, R.L. and John-Alder, H.B. (1994). Jumping performance of hylid frogs measured with high-speed cine film. *Journal of Experimental Biology*, **188**, 131–41.

Marsh, R.L., Olson, J.M. and Guzik, S.K. (1992). Mechanical performance of scallop adductor muscle during swimming. *Nature*, **357**, 411–13.

Milne-Thomson, L.M. (1966). *Theoretical aerodynamics*. Macmillan, New York.

Minetti, A.E., Ardigo, L.P. and Saibene, F. (1994). Mechanical determinants of the minimum energy cost of gradient running in humans. *Journal of Experimental Biology*, **195**, 211–25.

Mitchison, T.J. and Cramer, L.P. (1996). Actin-based cell motility and cell locomotion. *Cell*, **84**, 371–9.

Norberg, U.M. (1990). Vertebrate flight. In *Zoophysiology*, Vol. 27, p. 291. Springer-Verlag, New York.

O'Dor, R.K. and Webber, D.M. (1991). Invertebrate athletes: trade-offs between transport efficiency and power density in cephalopod evolution. *Journal of Experimental Biology*, **160**, 93–112.

Oster, G.F. and Perelson, A.S. (1987). The physics of cell motility. *Journal of Cell Science*, **8** (Supplement), 35–54.

Oster, G.F., Peskin, C. S. and Odell, G. M. (1993). Cellular motions and thermal fluctuations: the Brownian ratchet. *Biophysical Journal*, **65**, 316–24.

Parson, P.E. and Taylor, C.R. (1977). Energetics of brachiation versus walking: a comparison of suspended and an inverted pendulum mechanism. *Physiological Zoology*, **50**, 182–9.

Pearson, K.G. (1976). The control of walking. *Scientific American*, **235**, 72–86.

Pennycuick, C.J. (1975). On the running of the gnu (*Connochaetes taurinus*) and other animals. *Journal of Experimental Biology*, **63**, 775–99.

Peplowski, M.M. and Marsh, R.L. (1997). Work and power output in the hindlimb muscles of Cuban tree frogs *Osteopilus septentrionalis* during jumping. *Journal of Experimental Biology*, **200**, 2861–70.

Preuschoft, H. and Demes, B. (1984). Biomechanics of brachiation. In *The lesser apes: evolutionary and behavioral biology* (ed. H. Presuschoft, D. J. Chivers, W.Y. Brockelman and N. Creel), pp. 96–118. Edinburgh University Press.

Raibert, M.H. (1986). *Legged robots that balance*. MIT Press, Cambridge, MA.

Rayner, J.M.V. (1985). Bounding and undulating flight in birds. *Journal of Theoretical Biology*, **117**, 47–77.

Rayner, J.M.V., Jones, G. and Thomas, A. (1986). Vortex flow visualizations reveal change in upstroke function with flight speed in bats. *Nature*, **321**, 162–4.

Roberts, T.J., Marsh, R.L., Weyand, P.G. and Taylor, C.R. (1997). Muscular force in running turkeys: the economy of minimizing work. *Science*, **275**, 1113–15.

Rome, L.C., Loughna, P.T. and Goldspink, G. (1984). Muscle fiber activity in carp as a function of swimming speed and muscle temperature. *American Journal of Physiology*, **247**, R272–9.

Rome, L.C., Swank, O. and Corda, D. (1993). How fish power swimming. *Science*, **261**, 340–43.

Rosser, B.W.C. and George, J.C. (1986). The avian pectoralis: histochemical characterization and distribution of muscle fiber types. *Canadian Journal of Zoology*, **64**, 1174–85.

Russell, A.P. (1975). A contribution to the functional anatomy of the foot of the tokay, *Gekko gecko* (Reptilia, Gekkonidae). *Journal of Zoology*, **176**, 437–76.

Schmidt-Nielsen, K. (1972). Locomotion: energy cost of swimming, flying, and running. *Science*, **177**, 222–7.

Schmitt, D. (1999). Compliant walking in primates. *Journal of Zoology*, **248**, 149–60.

Shadwick, R.E., Katz, S.L., Korsmeyer, K.E., Knower, T. and Covell, J.W. (1999). Muscle dynamics in skipjack tuna: timing of red muscle shortening in relation to activation and body curvature during steady swimming. *Journal of Experimental Biology*, **202**, 2139–50.

Sheetz, M.P., Wayne, D.B. and Pearlman, A.L. (1992). Extension of filopodia by motor-dependent actin assembly. *Cell Motility and the Cytoskeleton*, **22**, 160–9.

Sleigh, M.A. and Blake, J.R. (1977). Methods of ciliary propulsion and their size limitations. In *Scale effects in animal locomotion* (ed. T. J. Pedley), pp. 185–201. Academic Press, London.

Spedding, G.R. (1987). The wake of a kestrel (*Falco tinnuculus*) in flapping flight. *Journal of Experimental Biology*, **127**, 59–78.

Stossel, T.P. (1993). On the crawling of animal cells. *Science*, **260**, 1086–94.

Stossel, T.P. (1994). The machinery of cell crawling. *Scientific American*, **9**, 54–63.

Suter, R.B. and Wildman, H. (1999). Locomotion on the water surface: hydrodynamic constraints on rowing velocity require a gait change. *Journal of Experimental Biology*, **202**, 2771–85.

Suter, R.B., Rosenberg, O., Loeb, S., Wildman, H. and Long, J.H. (1997). Locomotion on the water surface: propulsive mechanisms of the fisher spider *Dolomedes triton*. *Journal of Experimental Biology*, **200**, 2523–38.

Swartz, S.M. (1989). Pendular mechanics and the kinematics and energetics of brachiating locomotion. *International Journal of Primatology*, **10**, 387–418.

Swartz, S.M., Bennett, M.B. and Carrier, D.R. (1992). Wing bone stresses in free flying bats and the evolution of skeletal design for flight. *Nature*, **359**, 726–9.

Taylor, C.R. (1994). Relating mechanics and energetics during exercise. *Advances in Veterinary Science and Comparative Medicine*, **38A**, 181–215.

Taylor, C.R., Heglund, N.C. and Maloiy, G.M.O. (1982). Energetics and mechanics of terrestrial locomotion. I: Metabolic energy consumption as function of speed and size in birds and mammals. *Journal of Experimental Biology*, **97**, 1–21.

Theriot, J.A. and Mitchison, T.J. (1991). Actin microfilament dynamics in locomoting cells. *Nature*, **352**, 126–31.

Ting, L.H., Blickhan, R. and Full, R.J. (1994). Dynamic and static stability of hexapedal runners. *Journal of Experimental Biology*, **197**, 251–69.

Tobalske, B. (1996). Scaling of muscle composition, wing morphology, and intermittent flight behavior in woodpeckers. *Auk*, **113**, 151–77.

Tobalske, B.W. and Dial, K.P. (1994). Neuromuscular control and kinematics of intermittent flight in budgerigars (*Melopsittacus undulatus*). *Journal of Experimental Biology*, **187**, 1–18.

Tobalske, B. and Dial, K.P. (1996). Flight kinematics of black-billed magpies and pigeons over a wide range of speeds. *Journal of Experimental Biology*, **199**, 263–80.

Tobalske, B.W., Peacock, W.L. and Dial, K.P. (1999). Kinematics of flap-bounding flight in the zebra finch over a wide range of speeds. *Journal of Experimental Biology*, **202**, 1725–39.

Tobalske, B.W., Hedrick, D.L., Dial, K.P. and Biewener, A.A. (2003). Comparative power curves in bird flight. *Nature*, **421**, 363–66.

Videler, J.J. (1993). *Fish swimming*. Chapman & Hall, London.

Vogel, S. (1994). *Life in moving fluids: the physical biology of flow*. Princeton University Press.

Wagner, H. (1986). Flight performance and visual control of flight of the free-flying housefly. *Philosophical Transactions of the Royal Society of London, Series B*, **312**, 527–51.

Wainwright, S.A., Biggs, W.D., Currey, J.D. and Gosline, J.M. (1976). *Mechanical design in organisms*. Arnold, London.

Wainwright, S.A., Vosburgh, F. and Hebrank, J.H. (1978). Shark skin: function in locomotion. *Science*, **202**, 747–9.

Warrick, D.R. and Dial, K.P. (1998). Kinematic, aerodynamic and anatomical mechanisms in the slow, maneuvering flight of pigeons. *Journal of Experimental Biology*, **201**, 655–72.

Wasserthal, L.T. (1975). The role of butterfly wings in regulation of body temperature. *Journal of Insect Physiology*, **21**, 1921–30.

Webb, P.W. (1982). Locomotor patterns in the evolution of actinopterygian fishes. *American Zoologist*, **22**, 329–42.

Weibel, E.R., Taylor, C.R., Gehr, P., Hoppeler, H., Mathieu, O. and Maloiy, G.M.O. (1981). Design of the mammalian respiratory system. IX: Functional and structural limits for oxygen flow. *Respiration Physiology*, **44**, 151–64.

Weinstein, R.B. and Full, R.J. (2000). Intermittent locomotor behavior alters total work. In *Biomechanics of animal behavior* (ed. P. Domenici and R.W. Blake), pp. 33–48. BIOS, Oxford.

Weis-Fogh, T. (1973). Quick estimates of flight fitness in hovering animals, including novel mechanisms for lift production. *Journal of Experimental Biology*, **59**, 169–230.

Westneat, M.W. and Wainwright, S.A. (2001). Mechanical design of tunas: muscle, tendon, and bone. In *Tuna: physiology, ecology and evolution* (ed. B. Block and D. Stevens), pp. 271–311. Academic Press.

Wilga, C.D. and Lauder, G.V. (2000). Three-dimensional kinematics and wake structure of the pectoral fins during locomotion in leopard sharks *Triakis semifasciata*. *Journal of Experimental Biology*, **203**, 2261–78.

Williams, T.M. (1999). The evolution of cost efficient swimming in marine mammals: limits to energetic optimization. *Philosophical Transactions of the Royal Society of London*, **354**, 193–201.

Williams, T.M., Davis, R.W., Fuiman, L.A., Francis, J., Le Boeuf, B.J., Horning, M., Calambokidis, J., Croll, D.A. (2000). Sink or Swim: Strategies for cost-efficient diving by marine mammals. *Science*, **288**, 133–60.

Wilson, A.M., McGuigan, M.P., Su, A. and van den Bogert, A.J. (2001). Horses damp the spring in their step. *Nature*, **414**, 895–9.

Winter, D.A. (1990). *Biomechanics and motor control of human movement*. Wiley, New York.

Zug, G.R. (1978). Anuran locomotion—structure and function. 2: Jumping performance of semiaquatic, terrestrial, and arboreal frogs. *Smithsonian Contributions to Zoology*, **276**, 1–31.

Index

Made in the USA
Middletown, DE
31 January 2017